# BIRDS!
## From the Inside Out

Fourth Edition, Sixth Printing, 2010

# BIRDS!
## From the Inside Out

Written by Dan Gleason

Illustration and Design by Barbara Gleason

Photos by Dan Gleason

CraneDance Publications
a wing of
BGleason Design & Illustration, LLC
Eugene, Oregon

All written materials, illustrations and photographs
© 1999-20010 by Dan and Barbara Gleason

Cover Photo: Tree Swallow (*Tachycineta bicolor*)
by Dan Gleason © 2008

All Rights Reserved.

This book is protected by International copyright.

No part of this book may be reproduced or transmitted, in whole or in part, including illustrations, in any form, by any means, electronic or mechanical, including photocopying, scanning, digitizing, or recording, nor may it be utilized or stored in any information storage and retrieval system without advance written permission from the copyright owners, Dan and Barbara Gleason.

For information about reproducing any portions of this book, contact:
Barbara and Dan Gleason
CraneDance Publications
a wing of BGleason Design & Illustration, LLC
P.O. Box 50535, Eugene, Oregon 97405 USA
Phone/Fax (541) 345-3974
E-mail: info@bgleasondesign.com
http://www.cranedance.com

— • —

Library of Congress Cataloging-in-Publication Data

Gleason, Daniel and Barbara

Birds! From the Inside Out, by Dan Gleason, illustration and design by Barbara Gleason / Fourth Edition, Sixth Printing, 2010, p. 180

ISBN 978-0-9708895-0-8

| 1. | Ornithology—United States. | I. Title |
| 2. | Birds—Biology, United States. | |
| 3. | Biology—Birds, Avian. | |
| 4. | Science—Biology. | |
| 5. | Birdwatching—Birding. | |

# BIRDS!
## From the Inside Out

## Contents

Introduction ..................................................................... i.

CHAPTER

1  Feathers .................................................................... 1
2  Flight ...................................................................... 11
3  Avian Skeleton System ........................................... 21
4  Circulation and Respiration ................................... 33
5  Feeding Adaptations and Food Gathering ............. 43
6  Feeding Behaviors .................................................. 53
7  Bird Senses ............................................................. 59
8  Avian Mating Systems ............................................ 75
9  Brood Parasitism .................................................... 83
10 Hatching and Care of Young .................................. 91
11 Migration and Wintering ........................................ 99
12 Attracting and Feeding Wild Birds ....................... 109
13 Characteristics: North American Bird Families .... 115

APPENDICES

    Dictionary of Ornithological Terms ...................... 137
    Useful References .................................................. 156
    Index ...................................................................... 159

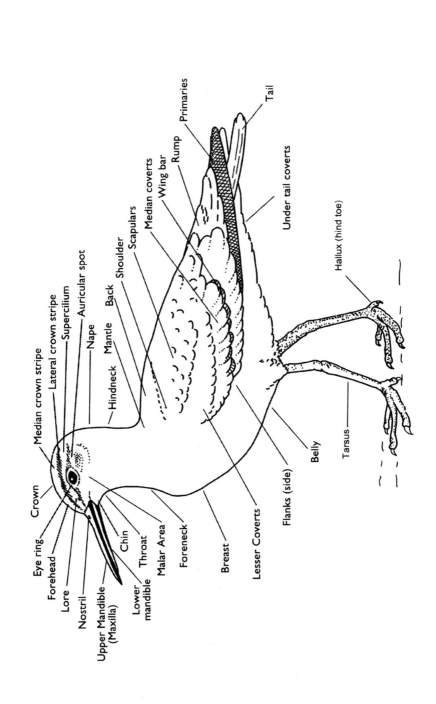

# Acknowledgements

For many years, we have led bird walks and field trips in and around Lane County, Oregon. We love watching new birders' eyes light up as they first identify a bird that is new to them. When student birders would want more information, we direct them to resources we knew of. But, many ornithology texts are far too unwieldy for busy adults to get through easily. To help such birders learn more, we taught a series of classes from 1995-99 and created illustrated handouts.

Then, we were contacted by a fellow ornithology instructor, who wanted to use our class materials in his classes. The first edition of this book was released in 1999, with later editions in 2000, 2005 and 2007. For this fourth edition we updated information, appendices, illustrations and the index to improve readers' access.

Over time, the book's use spread to many educational settings: community college ornithology, ecology and marine biology classes, Audubon Center master classes, senior center and nature center bird-oriented classes, and classes taught through bird-related stores.

This book would never have "fledged" except for some strong support. Two Oregon bird experts read the second edition and gave us valuable comments: Louise Shimmel of the Cascade Raptor Center is an eagle-eyed proofreader as well as a highly skilled avian rehabilitation expert, and her suggestions on many topics were invaluable. Reid Freeman, an expert birder then in Eugene, whose eyes spot field marks and written errors equally well, not only proofread but posed helpful questions about natural history and ornithology. Two additional non-bird-oriented colleagues, Ally LeCaux and Jerry Brown, helped as well. We knew if they had trouble understanding something, we had to change it!

We also want to thank our "Birds! From the Inside Out" students. Their great questions, and our answers helped create chapters. Here are even more notes than you got in class!

We hope the information we offer here adds much to the pleasure of your birding, allowing you greater appreciation of some of the amazing adaptations of birds.

And please realize that every individual CAN make a difference. Please be alert to dangers in your area to the birds we so enjoy. Habitats are disappearing at a rapid rate, so try to find a way to make a difference in your community.

*– Barbara and Dan Gleason*

# Introduction

We earthbound humans have always had a special fascination with birds. Their ability to fly, liberating them from the constraints of an entirely terrestrial existence, may be the greatest cause of this fascination. But we are also taken with their magnificent colors and plumage displays, and we are enraptured by the beauty and diversity of their many voices. We have loved and worshipped them, incorporated them into our mythology and attributed to them special, even magical powers. We have feared them and assumed they had malevolent powers when they were mysterious and unknown to us, as owls have often been considered as foretellers of death in numerous cultures. In our modern society, we have loved them and felt thrilled and, perhaps privileged, to study, watch and make birds a special part of our lives.

## What Makes Birds Unique?

There are more species of birds than of any other vertebrate class of animals: over 9,600 species are found throughout the world. Living birds range in size from the tiny Scintillant Hummingbird, *Selasphorus scintilla*, of Jamaica (2.3 g or about 5/1000 lb.; a little less than the weight of a dime), to the Common Ostrich, *Struthio camelus*, (150,000 g. or over 300 lb.). The largest birds to ever live were the extinct Elephant Birds, *Aepyornis*, of Madagascar. These giants were flightless and some weighed over 450 Kg (over 1,000 lb.), stood over 10 feet tall and laid eggs that had a capacity of nearly two gallons!

As different as birds are from one another, all have feathers, even flightless birds, something we recognize as being unique to birds. Feathers provide insulation and allow the possibility of flight. In addition to feathers, all birds have a beak, wings, and a unique avian-style foot. All are warm-blooded and all lay hard-shelled eggs. The toothless beak, covered with a horny sheath, is also unique to birds. (The beak-like snouts of the duck-billed platypus and echidna [spiny anteater] may be the closest approximation of birds' beaks found in other animals. These strange Australian mammals are unlike any other mammal in that they have a beak-like snout, lay eggs, and the females exude milk from numerous pores on their ventral surface rather than from discrete, paired teats as found in other mammals.) All birds have wings, but these are not unique avian structures — most kinds of insects and all bats have wings and fly. However, the wings of insects, bats and birds are all structurally different from each other. Later chapters will discuss many of these features in more detail.

## Introduction

# The Distribution of Birds

Birds are found throughout the world in nearly every habitat. They are among the most successful of all terrestrial vertebrates. Birds, like mammals, must breathe air, and therefore there are no birds that live in a completely submerged or benthic (ocean bottom) environment. Some seabirds dive to depths of over 200 feet, but they must return to the surface to breathe. Apart from the deep ocean or the vast interior of Antarctica, birds are found in most environments, even many extreme habitats. For instance, Sandgrouse nest on the hot Sahara sands, penguins and skuas live over the coastal Antarctic ice, and albatrosses land only to breed. The Andean Hillstar, *Oreotrochilus chimborazo*, a hummingbird in Peru, spends all of its life at an elevation of over 4500 meters (14,760 ft. ) and in Tibet, the Alpine Chough, *Pyrrhocorax graculus*, may be found at altitudes of 8229 meters (27,000 ft.)!

Given the mobility of birds, one might expect to find many of the same species throughout much of the world. Osprey, *Pandion haliaetus*, are found worldwide except in Antarctica and New Zealand. Mallards, *Anas platyrhynchos*, and Northern Pintails, *Anas acuta*, are found throughout the northern hemisphere. But, in the whole of the avian community, birds such as these are exceptions. Most species are much more limited in range and many have very limited distributions. Whole families are often represented in only one geographic region of the world. For instance, there are 63 species of wrens (Family Troglodytidae) found from North America through South America, yet, only one species, the Winter Wren, *Troglodytes troglodytes*, is also found in Eurasia. Over 300 species of hummingbirds are known. Most are found in South and Central America, but a few range as far north as North America, and the Rufous Hummingbird, *Selasphorus rufus*, even reaches into southern Alaska. The tropical regions of Africa and Asia would seem at first glance to provide comparable habitats, yet there are no hummingbirds found outside of the Americas. The Horned lark, *Eremophila alpestris*, is the only lark (Family Alaudidae) native to the New World, but over 75 species of larks are found in Europe and Asia. (The Eastern and Western Meadowlarks, *Sturnella magna* and *Sturnella neglecta*, are blackbirds [Family Icteridae], not true larks, and the Sky Lark, *Aluda arvensis*, was introduced into North America).

The greatest variety of birds is found in the tropics. The neotropics (South and Central America) are home to over 3300 native species of birds. With over 1/3 of all world bird species, this region is the most productive region for

## Introduction

bird life anywhere in the world. The species found here represent some 95 families, 31 of which are found nowhere else. In North America (technically speaking: north of the tropics of northern Mexico) there are over 750 species of birds known to breed. The Seventh Edition of the A.O.U. Check-list (published in mid-1998) covers the Americas north of the Panama Canal, including the Hawaiian Islands and the Caribbean Islands. It contains 2,008 species! Clearly, birds represent a very diverse group of animals in the world. As such, they have generated considerable study, from detailed scientific studies to the casual backyard bird-watcher.

## Birding as a Recreational Activity

Bird-watching, or "birding," is one of the most popular of all leisure activities, enjoyed by millions of Americans of all ages from all walks of life. In 2006, the USFW reported that 71 million people over the age of 16 participated in bird and wildlife-watching. That's one third of the 2006 U.S. population! The number of wildlife-watchers in 2006 totaled more than four times the attendance of all 2006 NFL football game attendees and wildlife-watchers spent more than 45.7 billion dollars. Wildlife-watching accounted for more than 1 million jobs, provided over 8.8 billion dollars in federal tax revenue and over 9.3 billion dollars in state and local taxes. Like any activity enjoyed by many people, there are many differences in the participants and the levels at which they choose to be involved.

For millions of people across North America, their birding experience consists mainly of putting up and maintaining bird feeders or bird houses. After an interest in birds in their yard has grown, novice birders often purchase a field guide, eager to find other birds in their neighborhood and local parks. A great many birders spend time in the field on a regular basis — weekly, sometimes daily, observing the habits and changes in the bird populations in their towns, and gradually gaining experience and expertise as their hours of activity increase. In many towns and cities there are regular field trips held by Audubon chapters or other birding groups, where novices can gain expertise as they observe and learn from the leaders. As people gain experience, they may develop particular expertise in a group of birds, or they may choose further study in the types of activities birds engage in. They may choose to pursue this interest through volunteer work with rehabilitation groups or banding experts, and they sometimes find that leading groups offers them opportunities to nurture other birders. Some experienced birders focus on building big lists of

## Introduction

birds, and some "chase" (travel to try to see) rare and unusual birds reported both close to home or at great distances. Some of these people travel thousands of miles each year in search of rarities. Occasionally, we might find a birder that is not at all interested in looking at robins because they are commonplace and easily seen. But others recognize that even a bird as common as a robin can still hold our fascination and teach us much about bird behavior and how they live.

Learning to identify as many birds as you can is fun and, for many, a reward in itself. But many of us want to know much more about birds. For such individuals, the greater their understanding of birds, the more satisfaction they derive from observations of bird activities. A better understanding of birds can also lead to a greater understanding and appreciation of the environment we share with them.

If you're about to embark on a study of birds or are seeking answers for questions that arise from your observations, we trust this book will aid in your efforts. Find a good field guide, grab your binoculars and head out to learn about the birds around you. It is our hope that this book will serve as a supplement to your chosen guide and provide you with some information and new ideas that are beyond the scope of an identification guide. This book is also a work in progress. There is always something more to learn — that is part of the excitement of birding. In future editions, we plan to add more chapters and expand upon existing material. In the meantime, we hope that we can satisfy some of your curiosity about birds and whet your appetite so that you will want to learn more about birds and the world we share.

*Dan and Barbara Gleason*

# Chapter 1
# Feathers

Perhaps the most obvious anatomical feature of a bird is its feathers. All birds, including the many flightless birds, have feathers. No other animals possess feathers. Feathers are mostly composed of **keratin**. This largely inert substance is made up of microscopic fibers in a matrix of proteins and is extremely durable. There are different types of keratin, some of which make hair, claws, nails and animal scales, but the keratin that makes feathers is unique in its molecular structure.

The arrangement and number of feathers on an individual bird varies between species, but birds

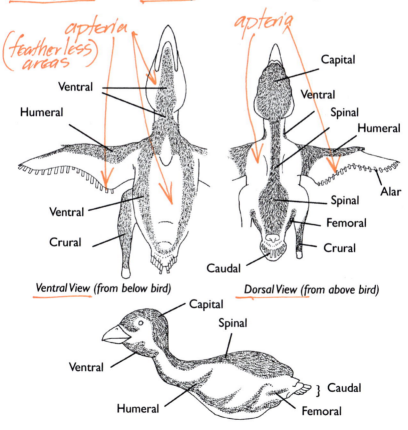

Figure 1-1. Feather tracts, or pterylae, of a typical songbird

**Birds! From the Inside Out**

## Chapter 1 – Feathers

Birds of a particular species all have the same number of feathers. Over the years, many people have counted the total number of feathers on various birds. These counts range from a low of 940 feathers on a Ruby-throated Hummingbird to a high of 25,216 on a Tundra Swan. Most songbirds have 1500 - 4000 feathers. The greatest concentration of feathers on most birds is around the head and neck. For most songbirds, this amounts to 30-40% of the total feathers.

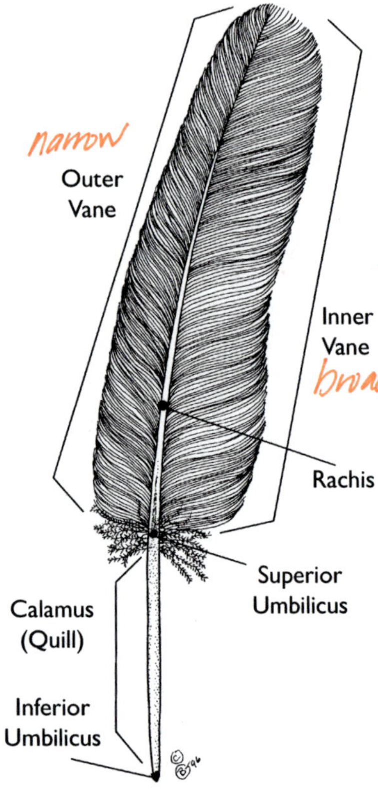

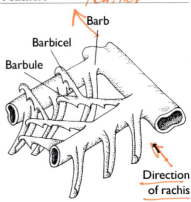

Figure 1-3. Structure of feather barbs, barbules and barbicels

Whereas an individual feather seems light, the entire plumage is a considerable portion of a bird's total weight. In fact, in most birds, the total weight of the feathers alone is 2 - 3 times the weight of the skeleton. The feathers of a Bald Eagle, for example, weigh about 700 grams (~1.5 lb.), 17% of its total mass of 4082 grams

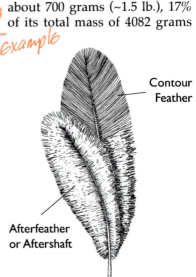

Figure 1-2. Flight feather, a contour feather found in a bird's outer wing.

Figure 1-4. Contour feather showing aftershaft or afterfeather.

(~9 lbs.). The skeleton, however, only weighs 272 grams (just over 1/2 lb.), or 7% of its mass — less than half of the weight of the feathers.

To look casually at a bird, it is easy to assume that the feathers grow uniformly over the body but, in fact, feathers grow only in distinct regions called ***pterylae*** or feather tracts. There are eight or nine major tracts which are further subdivided. The precise arrangement of these tracts and sub-tracts is one of many criteria used to distinguish and classify different species of birds. Between the pterylae are areas of bare skin known as ***apteria***. (Fig. 1-1). The apteria may assist with wing and leg movement, provide spaces for tucking legs and heads into the plumage, and they may facilitate heat loss (although this has not been fully established). Part of the evidence for assisting with heat loss comes from penguins, which have no apteria and a strong need to conserve heat.

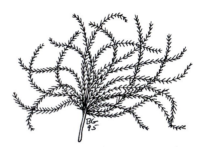

Figure 1-6. *Down feather*

## Vaned feathers - Contour (body) feathers, flight feathers

These are the most obvious feathers of a bird. All of the flight feathers of the wing and tail and most body feathers are vaned feathers. Vaned feathers have a long central shaft, called a ***rachis***, with a vane consisting of many barbs on either side of it. (Fig. 1-2). In flight feathers, this shaft divides the feather into two asymmetrical vanes. The narrow vane is on the leading edge of the feather and the broad vane is on the trailing edge.

***Barbs*** on the vanes of these feathers are divided into ***barbules***. The barbules which point toward the tip of the feather have tiny hooks which clasp the barbules pointing toward the base of the feather (Fig. 1-3). These hooks keep the barbs of the feather in a tightly interlocked condition, or ***pennaceous*** much like Velcro™. The barbs near the base are without hooks, soft and downy, termed ***plumulaceous***, to help trap air and provide insulation.

Figure 1-5. *Semiplume feather*

Figure 1-7. Bristle (left) and filoplume (right) are specialized feathers.

Many contour feathers have a small secondary shaft emerging from the base of the feather in the region of the rachis where the barbs begin to emerge (Fig. 1-4). These are termed *aftershafts* or *afterfeathers*. They are usually short and plumulaceous, and their primary function is to increase insulation. The afterfeather is usually much shorter than the main feather except in emus, where both the main feather and the afterfeather are nearly equal in length and structure.

The large pennaceous flight feathers of the wings are known as the *remiges*. The outer remiges are the large feathers attached to the hand and are called *primaries*. Although there is some variation in number of primaries between different species, the majority of birds have 10 primaries. These flight feathers are asymmetrical in structure with a narrow outer vane and a broad inner vane.

In further specialization, many of the barbs on the inner vanes of these feathers have small projections known as *friction barbs*, which rub against barbs of overlaying feathers. These friction barbs help to keep the feathers from slipping apart during flight.

The inner remiges, attached to the ulna, are called *secondaries*. The number of secondaries is much more variable. The short wing of the hummingbird may have as few as six, whereas the very long wings of shearwaters and albatrosses may have 30 - 40 secondaries.

The flight feathers of the tail are known as the *rectrices*. Most birds have 12 rectrices (snipe have 24 and anis and grouse have 18). Similar to flight feathers, some tail feathers are asymmetrical. The outermost feathers are the most asymmetrical, with the narrow vane on the side toward the outer edge of the tail. As one moves from the outer edge of the tail to the center, the feathers become more symmetrical and the four to six central feathers have symmetrical vanes. The rectrices help with some of the fine control of flight by aiding in steering and braking. The tail feathers of some birds are highly modified for use in courtship displays.

**Chapter 1 – Feathers**

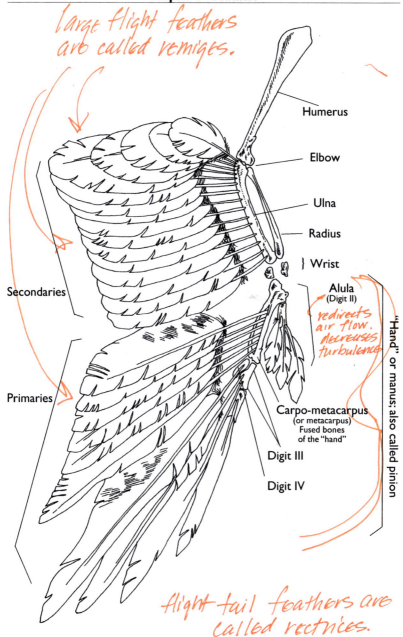

*Figure 1-8. Outstretched wing showing feather attachment.*

## Chapter 1 – Feathers

### Down feathers

While contour feathers provide some insulation, it is the down feathers (Fig. 1-6) that are most important for this function. Birds in the Arctic have a dense supply of down feathers; tropical birds generally have few. Soft down feathers (plumulaceous in structure) are found on both chicks and adult birds. Most down feathers have no central shaft or rachis, but in a few species, notably waterfowl, down feathers have shafts. All of the many plumulaceous barbs originate from the base. While the barbules are hookless, their soft texture enables them to become entangled with each other. This, and the fact that the down feathers lie close to the body (beneath the contour feathers), helps facilitate trapping air to provide insulation.

### Powder down

Powder down feathers are uniquely structured feathers which, unlike other feathers, continually grow and are never molted. The tips of the barbs constantly disintegrate into a very fine water-resistant powder. This powder helps waterproof and preserve the body feathers. They may be found throughout the plumage of many birds but occur in dense patches on the breast, belly and sides of herons and egrets.

### Semiplume

In structure, a semiplume (Fig. 1-5) looks something like a cross between a down feather and a contour feather. Semiplumes have a long central shaft but have soft, downy vanes. The rachis is always longer than the longest barb, in contrast to down feathers. Semiplumes enhance insulation and fill out the body's aerodynamic contour. Sometimes these feathers are large and serve an ornamental function, as in the plumes of an egret.

### Filoplumes

Filoplumes are hairlike feathers that have only a few barbs with barbules near the tip. These feathers help monitor movement and position of nearby vaned feathers. Movement of a filoplume stimulates a sensory cell at its base which signals the muscles to make adjustments. Filoplumes around flight feathers provide sensory input and help with small aerodynamic adjustments. Filoplumes along contour feathers may help monitor airspeed.

### Bristles

Bristles (Fig. 1-7) are short, stiff feathers with a long shaft and no barbs. (Some bristles may have a few short basal barbs). Bristles are found almost exclusively on the heads of birds. They provide some protection along with sensory input.

## Feather Care

A bird's very existence depends upon the care it gives its feathers. It is imperative that feathers are

*melanin: granular deposits*
*caratenoids: fat soluable pigments*

## Chapter 1 – Feathers

taken care of every day in order to ensure that they will continue to provide adequate insulation and remain structurally sound to allow flight. Feathers that are not cared for soon become brittle and easily frayed or broken, thus being of little benefit to the bird.

At the base of the tail, on the upper surface of the rump, is the ***uropygial gland*** or ***preen gland***. Secretions from this gland are rich in waxes, fatty acids, lipids (fats) and water. During preening a birds rubs its bill and head against the nipple of this gland, picking up the oily secretions which are then spread throughout all the feathers on the body by the head and bill. These secretions may help waterproof the feathers, but they mostly provide conditioning to help keep them supple. There are some additional benefits of these substances. Some of the fats help protect feathers against bacteria that digest keratin (the primary component of feathers). Other fats actually encourage the growth of certain fungi, which help protect the feather by making it difficult for feather lice to infect it.

## Feather Color

Feather colors are the result of a variety of pigments, the structure of the feather, or a combination of these factors. Pigments are laid down early in the development of a feather and become fixed as the feather matures and becomes keratinized. Since mature feathers are dead structures, there can be no change of color except early in feather growth (before it fully emerges from the feather follicle).

There are three main types of pigments found in birds: ***melanins***, ***carotenoids***, and ***porphyrins***. Melanins are granular deposits in the feathers (sometimes also in the skin, beaks or scales of the leg) whose surfaces absorb or reflect light, depending upon the wavelength of light. The smallest of these granules, called ***phaeomelanin***, are irregular in shape and are usually tan, reddish-brown or, occasionally, yellow. The larger granules, ***eumelanin***, tend to be more regular in shape and are dark brown, gray or black.

Melanins are found in all birds except total albinos, which are rather rare. In addition to providing color, the melanin provides other functions. Melanin gives feathers more strength and greater resistance to wear. Many birds, especially large birds, have darker primaries as additional insurance against feather breakage, i.e., the black wing tips of Snow Geese or American White Pelicans. Of course, there will always be exceptions, such as Great Egrets, but these exceptions are generally not long-distance migrants. Melanin also absorbs radiant energy. This may help to warm the bird and may play a role in helping to dry the feathers

## Chapter 1 – Feathers

(birds of wetter climates usually have darker colors).

Carotenoids are fat-soluble pigments which remain behind as the feather matures and becomes keratinized. These pigments (or the components for manufacturing them) are all derived from the diet of the bird. Carotenoids account for most yellows, oranges, reds and purples found in birds. Male House Finches that are able to find foods rich in carotenoids develop the rich, purplish-red color that is characteristic of this species. Males that are less able to find such food produce weaker reds or may sometimes be orange or yellow. An orange male that learns to find and eat carotene-rich foods will remain orange until the next molt, when his feathers will develop a richer red color (remember, mature feathers are dead structures and cannot change color). A series of experiments by Geoffery Hill has demonstrated not only this phenomenon but the fact that redder males are preferred by females. Presumably, redder males make better mates because they are finding better quality food sources. This directly benefits the female through ritualized courtship feeding and the feeding of her young during the breeding season. Orange or yellow males are less able to obtain mates.

Porphyrins are the final group of pigments found in some birds. These are chemicals related to hemoglobin or derived from breakdown products of hemoglobin. These generally produce bright brown tones and, in a few uncommon cases, some greens. Porphyrins tend to be unstable and begin to degrade in the presence of strong light.

Apart from pigment, colors in birds can result from the structure of the feathers themselves. For example, there are no blue pigments in birds, yet many birds appear blue in color. Blue feathers are only blue in reflected light. In transmitted light, the feather will appear gray. Try this with the next feather you find from a bluebird or jay. As you look down upon the feather, it will look blue. Now hold the feather up to the light and notice that the light coming through the feather makes the feather look gray. Tiny air vacuoles in the keratin surrounding the melanin granules scatter the blue light. The particular shade of blue is dependent upon which exact wavelengths are reflected or absorbed. A yellow pigment in this keratin layer may add to the scattered blue producing the green that is found in many parrots.

Color may also be produced by a combination of structural features and pigmentation. Most greens, such as in parrots, are the result of yellow pigments combined with a structure that reflects blue light. The combination of blue and yellow produces green. Red carot-

## Chapter 1 – Feathers

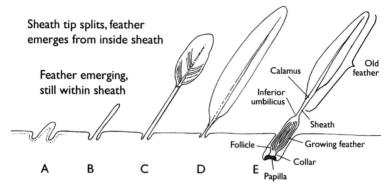

*Figure 1-9. Feather growth, shown in five steps.*

enoid pigment combined with structural blue results in violet or purple. A dull olive-green, like that of vireos, is usually a combination of melanin and carotenoids. If the melanin is in the barbs, the carotenoid is usually in the barbules and vice-versa.

*Figure 1-10. Three-week old Cockatoo showing newly emerging feathers. (Photo courtesy of Karen Ackoff)*

## Feather growth

As a feather develops and begins to grow, the bud emerges through the dermis (Fig. 1-9, A). The new feather is wrapped in a tube-like sheath (Fig. 1-9, B). The process of keratinization (feather "hardening") proceeds from the tip downward. The keratinized feather begins to expand (Fig. 1-9, C) and break away from the sheath. The feather is mature (Fig. 1-9, D) when it is completely elongated and expanded. During a molt, a new feather bud develops in the follicle at the base of the old feather (Fig. 1-9, E). As the new feather begins to grow and emerge, it eventually pushes out the old feather, which is shed as the new feather is ready to break away from its sheath.

**Chapter 1 – Feathers**

Carotenoids: obtained from diet. red, yellows...
Melanin: granules that produce reddish browns, browns, tan, rarely yellow.
Porphyrins: related to hemoglobin and produces bright browns.

Sometime pigments mix, sometimes reflected light and pigments.

# Chapter 2
# Flight

For countless generations, people have looked to the skies and envied the flight of birds. Flight has inspired hundreds of inventions, poems, essays and songs. The ability of a bird to fly is perhaps its most notable feature and many species exhibit what seems to be complete mastery of the air currents. Some birds are capable of hovering in one place, some dive at great speeds, albatrosses glide with few wing beats and may spend months or even years at sea before returning to land, and a few birds (hummingbirds and tropicbirds) are even capable of making short flights backwards! The perception of effortless flight serves well as a model of freedom, but is it really so effortless and how is it accomplished?

Flight in birds is not accomplished merely by flapping the wings up and down. The mechanics of avian flight are complex and have mystified human observers for many centuries. A failure to understand these complexities has led to the creation of many failed inventions intended to get humans off the ground, some with disastrous results. Modern aircraft use the same physical principles as birds do to achieve flight, but aircraft designers do not attempt to have their aircraft imitate all of the actions of birds associated with flight.

An understanding of both the structure of a bird's wing and the physical mechanics of flight are important for a complete understanding of bird flight. Unique structural and physiological adaptations are also necessary if a bird is to achieve flight.

## Keep Weight to a Minimum

Weight reduction is essential and birds accomplish this in several ways. Most flying birds have a thin, lightweight skeleton. In many birds, the weight of the skeleton is actually less than the total weight of all the feathers! The long bones are usually hollow and lack marrow, and there is a reduction in the overall number of bones as compared to other vertebrates. (See Chapter 3 on the skeleton for more details.) Birds lack teeth and the heavy jaw typical of other vertebrates. By laying eggs, female birds are free of the weight burden of internally developing young, and in most species, only one oviduct is functional. In males, often only one of the testes is functional. In addition, the reproductive system shrivels in the non-breeding season to a small fraction of its breeding season size, further reducing weight. The timing of this is an important consideration for long-distance migrants.

## Other Adaptations for Flight

Birds have other physiological adaptations that are important for flight. They are all homeothermic ("warm-blooded") and, like mammals, have a four-chambered heart that prevents oxygenated blood from mixing with unoxygenated blood, assuring that tissues are most efficiently bathed with a supply of oxygen. An efficient, fast-acting digestive system prevents birds from having the increased weight of partially digested food for long periods of time. Birds also eat an easily digestible, high-energy diet. Mammals reduce the toxicity of waste products by having a dilute urine, but doing so imposes an extra burden of weight– insignificant for most mammals, but detrimental for a flying bird. Instead of a large volume of dilute urine, birds excrete a uric acid undiluted with large volumes of liquid. Finally, birds have the most efficient respiratory system of all vertebrates, a system designed to maximize the rate of exchange of respiratory gasses.

## Basic Aerodynamics

A common misconception is that birds use their wings in the air much as we would use oars to row a boat. Looking at the slow, steady wing beats of a Great Blue Heron, it is easy to get this impression and mistakenly believe that it is pushing down and backwards against the air. In fact, this is not the action of a bird's wing nor does it explain how a bird stays in the air. A bird actually moves its wings down and forward through the air in such a manner that air flowing over the surface of the wings generates lift in the same way the fixed wings of an airplane moving through the air generate lift.

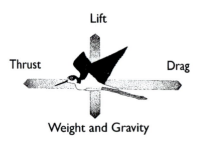

Figure 2-1. *The four opposing forces acting upon a bird: thrust, drag, lift and weight/gravity.*

Four basic forces act upon a bird moving through the air - two with a horizontal influence and two with a vertical influence. The horizontal forces are **thrust** (forward direction) and, acting in opposition to it, **drag** (rearward). The vertical forces are also opposites and are called **lift** (upward) and **weight** (downward). It is through the interaction of these four forces that a bird (or an airplane) is able to fly.

Imagine an airplane sitting on the end of the runway. While the plane is idle, lift, thrust and drag are all zero. The power of the engines is increased and the plane begins to move forward (thrust). The curved surfaces of the wings moving through the air generate lift. As thrust increases, so does lift. Eventually, lift exceeds the downward pull of gravity (weight) and the plane rises. As the angle of the plane's ascent increases, turbulence develops along the trailing edge of the wing creating frictional forces (drag) which oppose the forward movement gained by thrust. The pilot must take care that the angle is not too steep, for if drag exceeds thrust, lift will be decreased and the plane will fall.

The shape of a wing (bird or plane) in cross-section is not flat, but is the shape of an airfoil, that is, somewhat teardrop-shaped, with the upper surface having a greater curvature than the under surface. As a result, air moving over the upper surface must travel faster than the air moving over the lower surface. This creates lower air pressure above the wing than below, and an upward force, **lift**, is generated. This is known as the **Bernoulli Effect** in honor of Daniel Bernoulli (1700-1782), the Swiss physicist who first described this principle. He described this effect with water, not air, but the principle is the same for gasses as it is for liquids. He observed that water moving through a pipe has a certain pressure. If a portion of the pipe is constricted and smaller in diameter, the water moving through the constricted portion increases

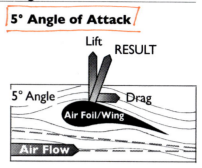

Figure 2-2. *As the angle of attack rises, lift increases, initially faster than drag. The lift/drag ratio reaches its maximum at an angle of 5°.*

in speed and the pressure drops. If the pipe is larger in diameter, the water flow decreases and the pressure rises. This constricted pipe is similar to the cambered (curved) surface of a wing. Air flowing over the upper surface is increased in speed and the pressure drops while the speed of air flowing under the lower surface is decreased and the pressure increases. The net result is lift. Most of the lift is generated by the decrease in pressure on the upper surface. You can test this effect for yourself. Using an ordinary sheet of notebook paper, hold it by the corners and allow it to droop down. Put the paper to your mouth and blow over the top (don't blow underneath the paper). The rapidly moving air that you have created over the top of the paper has a lower pressure than the static air on the underside of the paper, and, as you will notice, the drooping end of the paper rises.

## 15° Angle of Attack

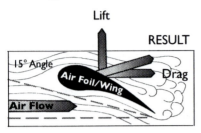

*Figure 2-3. When the angle of attack increases beyond 5°, drag increases more than lift. At a 15° angle, the airfoil (or wing) stalls because the flow of air over the upper surface becomes turbulent, which decreases the force of lift and accelerates the force of drag.*

Lift is not the only force generated as the wing moves forward. As lift increases, so does drag. At the trailing edge of the wing, the air below the wing tends to curl back over the top of the wing in an attempt to fill the lower pressure region above. This creates turbulence or drag and decreases the efficiency of the wing. The **angle of attack** is the angle at which the wing is held as it moves through the air. An angle of attack of 5° is the most efficient, creating the most lift. If the angle is increased to greater than 5°, the turbulence becomes stronger and drag increases. At an angle of approximately 15°, drag increases to a point where lift and the force of gravity (weight) are equal and the wing stalls. Beyond this, gravity overcomes lift and the bird or plane falls. A kestrel or kingfisher hovering in the air is maintaining that stall point and remains stationary. This requires a lot of work, and they can only hover for a few short moments.

Stalling is also important as a bird lands, for it must maintain a momentary balance between slowing and stopping its forward motion while grasping a perch, without falling from the sky. A steep angle of attack also allows for a quick takeoff. But how does a bird prevent stalling, which would be detrimental to getting into the air? Part of the answer lies with a structure on the wing called the **alula**. The alula is made up of the bones of the first digit on the manus ("hand") and the three small feathers that are attached to it. Together, they create a small forewing which redirects the flow of air over the top of the wing, decreasing the turbulence and increasing lift. Once in flight, the alula is usually pulled back down to become part of the smooth contour at the front of the wing. It is perhaps easiest to observe on a *Buteo* (such as a Red-tailed

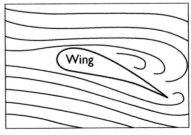

*Fig. 2-4. Airfoil/wing with high angle of attack produces turbulence along its upper surface; results: high drag and resulting loss of lift.*

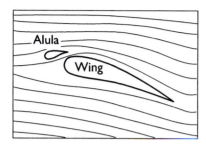

Fig. 2-5. *The alula redirects air over the upper surface of the wing, thereby reducing drag and creating greater lift, since, as an airfoil itself, it produces its own lift, adding to that of the wing.*

Hawk) as it soars on a thermal. The rapidly rising air increases turbulence over the back of the broad wing and the alula is lifted to smooth the airflow over the wing. But getting into the air requires more of a bird than simply elevating the alula.

The primary feathers are attached to the manus or "hand" portion of a bird's wing. As the wrist is flexed, this portion of the wing moves up and down a much greater distance than does the inner wing. This is especially important as a bird launches itself into the air. The greater downward distance helps to generate much of the lift necessary for takeoff. In addition, the primary feathers are spread at the wing tips and each acts as a separate airfoil generating its own lift. During this time, you could consider the primary feathers to be functioning like the blades of an airplane propeller. Once in the air, with increased thrust the inner wing now generates much of the lift. Removing a few of the primaries from each wing of a bird may disable it from flight because it cannot get into the air. Removing several secondaries (along the inner wing) does not usually prevent a bird from flying, but may decrease its efficiency in doing so.

There are two other properties that affect the ability of a bird to fly: **aspect ratio** and **wing-loading**. The aspect ratio is the ratio of wing length (**span**), measured from shoulder to wing tip, to the wing width (**chord**), measured front to back. Sailplanes have a high aspect ratio, i.e., long, narrow wings. Albatrosses and shearwaters also have wings with a high aspect ratio. With a long surface area, lift is easily achieved. The narrow wing produces little drag, and the pointed wing tips help balance the differences in pressure between upper and lower surfaces at the end of the wing where small vortices of air, creating turbulence, would otherwise be created. Such wings are highly efficient aerodynamically, but are somewhat weak structurally and cannot be flapped quickly for rapid flight nor can

*Albatross Wing Shape*

## Chapter 2 – Flight

they be used to make quick turns in the air. An albatross's wings are so long that, in still air, it cannot flap its wings sufficiently to become airborne. The deep wing beats necessary for take off are not possible because the wing tips would strike the ground before the movement could be completed. In moving air, the wing surface is so great that lift is quickly and easily achieved. An albatross simply faces into the wind, spreads its wings and is soon airborne. Lack of wind makes becoming airborne more difficult, so the albatross must do as a plane does: it spreads its wings and runs along a "runway" until sufficient speed is reached to create lift.

Wing-loading is the ratio of the body weight to the total wing area. A sailplane or albatross has low wing-loading and can fly at slow speeds. A bird (or plane) with high wing-loading must fly fast to stay airborne. Ducks have short, narrow wings with high wing-loading and so must fly fast (most ducks can easily fly 40-60 mph) to stay in the air. Swallows also have high wing-loading and must fly fast, but their short, narrow wings are a great advantage. They allow quick maneuvering in the air, an important consideration when trying to catch flying insects.

### Gliding and Soaring

Birds with high wing-loading, like ducks, must constantly flap their wings to stay airborne. Gliding is minimal and results in a rapid loss of altitude and airspeed. Many birds, however, can and do make use of gliding and soaring. Gliding is the simplest form of flying, requiring no effort on the part of the bird. Some other vertebrates, such as flying squirrels and some lizards, make use of gliding. It was undoubtedly the earliest form of flying used by avian ancestors. Gliding requires little effort and always results in a loss of both altitude and airspeed. The most efficient gliders descend only a small amount compared to the distance covered in a forward direction.

While gliding is a slow descent, soaring is the ability to maintain a constant altitude, or even increase that altitude, without flapping the wings. To be successful at soaring, a bird must be large and have low wing-loading. Small birds cannot soar because their small size does not give them enough momentum to remain unaffected by pockets of erratic air that might affect stability. Small birds also typically have higher wing-loading. The simplest method of soaring is to glide into rising air currents. This is called **static soaring**. Rising air currents may be caused by winds moving air against large obstacles such as cliffs and steep hills or by thermals, rising

*Buteo Wing Shape*

pockets of warm air. Birds such as hawks and vultures regularly take advantage of these rising air currents, which require little work by the bird to stay airborne. Turkey Vultures use static soaring as their primary method of flight and take to the air later in the morning when the air has been warmed and thermals have started. Birds that routinely depend upon static soaring typically have smaller breast muscles in proportion to their size than do birds that do not use static soaring, like songbirds or ducks.

Static soaring is only effective over land surfaces. Thermals and rising air pockets are seldom found over the ocean waters and, at best, are unstable when they do occur. Fortunately for albatrosses and other seabirds, other forces are available that allow the birds to engage in **dynamic soaring**. The trade winds and similar steady air currents are in constant motion over ocean surfaces. At the interface of air and water, friction slows the winds so that the wind speed within a few feet of the surface is slower than winds higher up. Albatrosses and other soaring seabirds use this velocity gradient to their advantage. This dynamic soaring involves gliding from a higher altitude (15-50 feet) to a height of just inches above the water. This is done in the direction of the wind. As the bird is in this gentle, downward glide, it picks up speed. When it is just above the waters' surface, it uses its momentum to turn into the wind and begin to rise. As it soars higher, the increased wind speed gives higher lift and the momentum gained by the previous descent carries the bird a long distance forward. Albatrosses are so efficient at this that they seldom need to flap and can travel more than a thousand miles each week using this method of flight.

## Ground Effect

Albatrosses and shearwaters also make use of another phenomenon known as **ground effect**, which occurs when the distance above the ground (or water) is less than the wing span. The air funneled below the wing reduces drag, flight becomes more effortless and slower speeds are possible. Albatrosses and shearwaters often fly long distances just inches above the water's surface. In fact, the name shearwater comes from this type of flight, as it appears that the wing tips are "shearing" the water's surface. These birds will periodically lift above the water just a few inches more to allow enough room for an occasional wing flap. Skimmers spend all of their foraging time just inches above the water and make use of this same phenomenon. Relatives of gulls and terns, skimmers have a rather bizarre-looking beak. Their lower mandible is larger and much longer than their upper mandible. As they fly over the surface of the water, they extend the lower mandible below

## Chapter 2 – Flight

the surface of the water. When contacting a fish, they snap the jaw closed, trapping the fish. The leading edge of the mandible is sharp to reduce resistance to both the air and the water. Skimmers also have long, slender wings. Their inner wing (from shoulder to wrist) is longer than that of its cousins, the terns. This long, narrow inner wing produces greater lift. On the upstroke, this long inner wing elevates the wrist higher, allowing the outer wing to have room above the surface for a deeper stroke to gain more thrust and lift. Rough waters prevent a bird from flying only inches above the surface and taking advantage of this ground effect. Skimmers, who depend upon this as a way of life, are thus limited in range to sheltered bays and lagoons where the waters remain calm.

### Flapping Flight

Flapping flight is not as simple as moving the wings up and down. At the beginning of the downstroke, the wing is fully extended. The wing is pulled down and forward to generate lift. During the downstroke, the flight feathers are held together to help force air over rather than through the wing. On large, broad-winged birds like eagles and vultures, the primaries at the wing tip may be spread, forming slots which allow each primary to act like a propeller blade and provide additional lift. The rachis of each flight feather is off-center,

*Figure 2-6. Black Skimmer in feeding flight.*

with the narrower vane forming the leading edge of the feather. As the wing moves up and down, the feathers rotate along the axis of the rachis. On the downstroke, this action helps draw the feathers tight against each other. Additionally, the outer portions of some barbules on the flight feather have a raised, or lobed, barbicel. This provides friction against the vane of an overlapping feather, helping to keep all of the feathers together.

At the bottom of the downstroke, the wing is folded partially in, decreasing the surface for less air resistance during the next upstroke. As the wing is raised, the pressure against the flight feathers causes them to rotate slightly in the opposite direction from the rotation during the downstroke. This allows the feathers to pull

## Chapter 2 – Flight

apart from each other slightly and allows air to pass through the wing area, further decreasing the resistance. At the top of the upstroke, the wing is once again fully extended. As the wing is extended, the outer wing (hand) moves outward and spreads the flight feathers.

Hummingbirds have a unique wing structure with increased flexibility that allows hovering flight while feeding. Unlike kestrels, osprey, kingfishers and other hovering birds, hummingbirds are not creating a stall condition which requires large expenditures of energy. Hummingbirds have a very short inner wing; most of the wing that you see is the outer wing. The wing rotates at the shoulder, turning the wing completely over so that the front of the wing is the leading edge on both the downstroke and upstroke. (Remember that a hummingbird, while hovering, holds its body nearly vertical so that the wing's downstroke is forward and the upstroke is rearward.) The wing is acting as an airfoil, creating lift throughout the entire range of each wing beat. Refined movements of the wing, tail and body all determine the balance of forces that result in hovering, forward, or even backward flight.

The leading edges of birds' inner wings is formed by membranes and skin over a tendon, the **patagium**, not over bone. When the wing is fully extended the

*Figure 2-7. Hummingbird, in hovering flight*

bones are not straight, but rather the elbow is still bent and pointing rearward. The patagium is a tendon which attaches the wrist to the coracoid bone at the shoulder. As the wing is opened, this tendon is stretched and pulls against the wrist to help open the hand. This can be demonstrated on a captive bird. As you pull the wing out, you will notice that the hand, forming the outer wing, is automatically extended.

### Tails and Wing Shapes

While flight is achieved through the action of the wings, fine control, especially braking and steering, is assisted by actions of the tail feathers. A bird that has lost its tail feathers is perfectly capable of flying but lacks subtle control, making maneuvering and landing more difficult. Long, wide tails provide more surface area, increasing lift and decreasing overall wing-loading, which enables the bird to have more stability at slower speeds.

Long tails also offer the benefit of the formation of a longer surface over which air must flow

## Chapter 2 – Flight

*Accipiter Wing Shape*

as it moves over the wings and back, and the tail forms an aerodynamic "slot". With this additional slot formed by the tail, less turbulence is created behind the wing, assisting with lift, especially at the high angles of attack created during takeoff and landing.

Birds with short tails, such as loons, pelicans and ducks, have feet that extend beyond the end of the tail. They spread these feet to assist in braking, as a spread tail does for other birds.

Wing shape and tail length put some limits on the types of habitats available to particular birds. Birds of open country typically have longer and somewhat more pointed wings, allowing them to soar above meadows and fields. Birds living and gathering food in heavier vegetation require shorter, more rounded wings and longer tails to help them maneuver between trees and shrubs. Accipiters and jays are woodland birds having such a wing/tail pattern. Small songbirds have shorter, rounded wings which enable them to seek quick shelter in dense shrubs. Loons, with their very short tails and short, pointed wings, lack the ability to take off abruptly or land in short distances. Consequently, loons are only found on large, open bodies of water.

Flight enables birds the opportunity to use a wider variety of habitats than are available to most vertebrates. While we often admire the song or rich colors of a bird, flight is perhaps the ability that we most envy.

# Chapter 3
# Avian Skeletal System

Long ago, the ancestors of modern birds overcame many of the constraints of terrestrial living and took to the air. Birds and bats are the only vertebrates to have achieved true flight. Feathers are the most visible adaptation in birds but both birds and bats have highly modified, yet different, skeletal systems. The earliest birds, like *Archaeopteryx*, retained many reptilian features including a long, bony tail and a full set of reptile-like teeth. Modern birds have lost these characteristics but still share many characteristics with modern reptiles. The skeletal system is one area where these characteristics are similar.

Birds have many adaptations to reduce their weight in order to increase flight efficiency, including the skeletal system's light weight. In frigatebirds, for example, the weight of the skeleton is less than the total weight of all its feathers! Birds have fewer bones in many portions of the skeleton compared to other vertebrates and many bones have become fused together over the course of evolution. For the most part, bird bones are thin, lightweight and often **pneumaticized**, or hollow. This is especially true of the long bones of the limbs, in marked contrast to mammals whose bones are filled with marrow.

While most birds have hollow bones, there are exceptions to this generalization. The ratites (kiwis, ostrich, emu, etc.) and most other non-flying birds do not have hollow bones. Some flying birds, such as loons and gulls also lack hollow bones.

The pneumaticized bones are often reinforced internally with strut-like projections which, like cross beams on a bridge, crisscross the hollow spaces

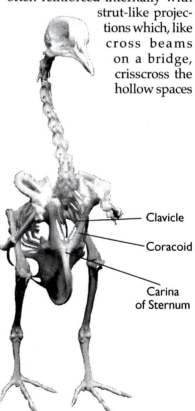

Figure 3-1. Avian skeleton (Rock Pigeon shown here)

## Chapter 3 – Avian Skeletal System

and give mechanical support. Air sacs of the respiratory system, especially around the pectoral region, often extend into these hollow spaces. These hollow bones are often described as being a weight-reducing mechanism. While weight-reduction is important for birds, some bones, such as the humerus, the leg bones and the sternum of the rib cage are actually heavier than those same bones in mammals of equivalent size.

### The Skull

Like most bones in a bird's body, the skull is relatively thin and lightweight. Prominent features of the avian skull are the large orbits or eye sockets. Birds typically have eyes much larger than mammals of equivalent size. The eyes of a Great Horned Owl, for example, are nearly equal to the size of the human eye. To help support such large eyes, birds have a ring of 8-12 bony plates located within the globe

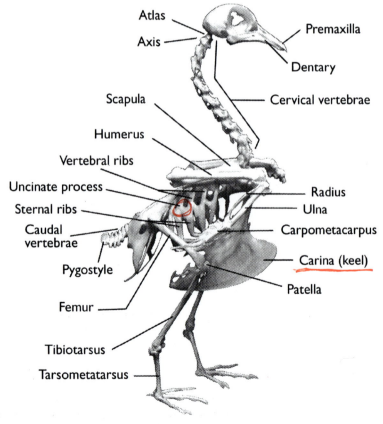

Figure 3-2. *Pigeon skeleton, side view*

## Chapter 3 – Avian Skeletal System

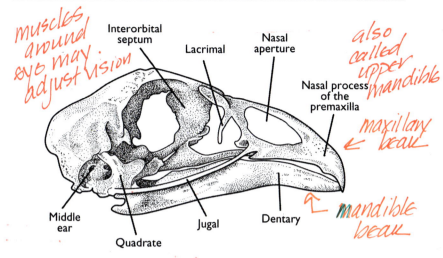

Figure 3-3. Chicken skull

of the eye. This **sclerotic ring** is a feature that birds share with their reptilian ancestors. Muscles compressing against this ring may also help change the shape of the eye, assisting it in its ability to accommodate for different viewing distances. Separating the two orbits is a very thin **septum**. When examining dry bird skulls, you may notice that the septum has been partially broken away, leaving an opening between the orbits. To accommodate the large eyes, the brain case is forced rearward behind the eyes.

The beak, a structure unique to birds, is another equally prominent feature of the avian skull. (The "beak" of the duck-billed platypus is anatomically different and only superficially resembles a bird's beak.) In life, the beak is covered with a horny sheath that may sometimes give the beak a very different size and look than if the shape were determined solely by the bony structure. For example, a puffin's beak during the breeding season has much greater depth (from top to bottom) because of enlargement of this horny sheath.

The beak is divided into two separate portions. The lower portion, or lower jaw, is called the **mandible** and the upper portion is called the **maxilla**. The term maxilla actually refers to a specific bone and the entire upper portion of the beak is, therefore, often referred to as the **upper mandible**. (Without "upper" as a modifier, the word mandible always refers to the lower jaw.)

The lower jaws of both birds and reptiles are made up of 5-6 individual bones that are fused into a

**Birds! From the Inside Out**     23

single element. In birds, the **dentary** bone is the most prominent and makes up the major portion of the lower jaw. The other bones of the mandible form the posterior region, but are fused so that they are difficult to distinguish from each other. In contrast, a mammal's jaw is derived from a single bone, the dentary. Modern birds' weight is reduced by not having teeth or the massive jaw necessary to support teeth. The beak, therefore, helps replace the teeth in function.

The mandible, instead of articulating directly against the skull as it does in mammals, articulates with paired bones, the right and left **quadrates**. The quadrates thus serve as an intermediary between the skull and the mandible. Both right and left quadrates also work against the right and left **zygomatic arches**, which facilitates flexion at the base of the upper jaw (maxilla) where it meets the skull. As a result, many birds have the ability to move the upper portion of the beak as well as simply moving the lower jaw, although to a much lesser degree.

Skimmers (see illustration of Skimmer, page 50) are an interesting exception to this pattern. Their mandible is rigid against the skull while the upper portion of the beak can open wide. Skimmers feed by flying just above the water with their mandible cutting the water's surface. If the lower jaw was not affixed to the skull, it could easily be torn loose from water resistance or from striking one of the fish upon which it feeds. Parrots and honeycreepers also have a high degree of mobility of the upper beak.

In addition to the mandible and the zygomatic arches, the quadrates also articulate against the **pterygoid** bones which are at the base of the eye orbits. The pterygoid bones connect to the **palatine** bones which form the **hard palate** in the roof of the mouth. The structure of this palate has often been used by taxonomists when attempting to classify birds into different orders and families.

Birds and reptiles have a single middle ear bone, the **columella**, unlike mammals which have three. The shape of this bone differs slightly between different orders of birds. This is another of the many structural features used by taxonomists to classify birds.

Located at the base of the skull is the **foramen magnum**. This is the large hole that is located atop the spinal column and through which the spinal cord runs to connect to the brain stem. Adjacent to the innermost side of this hole is a ball-like projection known as the **occipital condyle**. (A condyle is a rounded process on the part of a bone which articulates against another bone.) This condyle articulates with the first of the **cervical vertebrae** (neck bones)

## Chapter 3 – Avian Skeletal System

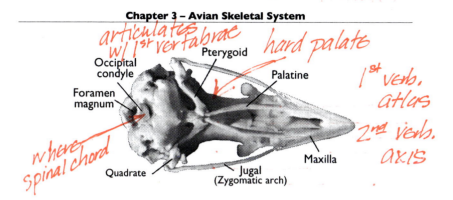

Figure 3-4. Chicken skull - from below

known as the **atlas**. The second vertebra is called the **axis**. If you look at a mammalian skull, you will notice that there are two occipital condyles. This allows the head to move up and down or to twist with respect to the atlas but limits side-to-side motion. (Look forward and tilt your head from side-to-side. Notice that you are moving all of the bones of your neck, not just your skull.) The single, ball-like condyle of a bird allows for much greater flexibility of movement. A bird often collects oil from the preen gland (located at the base of the tail) with the top of its head and spreads this oil throughout the plumage. The great flexibility of movement of the skull and neck help accomplish this task.

### The Bones of the Body

One of the most striking features of the avian skeleton is the large **carina** or keel protruding from the **sternum** of the rib cage. The broad, flat carina serves as a point of muscle attachment for the large flight muscles that raise and lower the wings. The muscle (**pectoralis major**) which pulls the wing down is the largest muscle in a bird's body and needs a large surface on which to attach. In some birds, the carina is not merely a broad, thin blade. In swans, for example, the carina is thickened and hollow in its innermost portion. Within this hollow is found a coil of a much-elongated trachea. The pectoralis major of flightless birds is not as massive as in flying birds, thus these birds lack a carina and have a flat sternum much as we do.

Woodpeckers, nuthatches and other birds which spend much of their time clinging to the trunk of a tree have a very shallow carina. A large, deep carina would interfere with their ability to get close to the tree trunk. These birds are strong fliers and require large flight muscles. These muscles must be spread across the entire rib cage since the carina itself is inadequate to provide the full support that they require.

The bones of the neck are called the **cervical vertebrae**. The number of cervical vertebrae is fixed within a species but is variable between species. In birds, the fewest number of these neck bones is 13 and the greatest number is 25 in some long-necked birds. In contrast, mammals have only seven cervical vertebrae. (The only mammalian exceptions to this are the manatee and two-toed sloth which each have six, the anteater with eight and the three-toed sloth which has nine.)

Directly posterior to the neck are the **thoracic vertebrae**. There are five of these bones which, in birds, are fused together giving greater stability and support during flight. A large **vertebral rib** extends from the sides of each of these vertebrae. The ends of each of these ribs articulate against the end of a **sternal rib** arising from the sternum. Together these form the rib cage. Midway along each vertebral rib is a small process, called the **uncinate process** which projects toward the back and overlaps the next rib to the rear. This helps provide stability against the forces generated by the muscles during flight. Without these processes, the thoracic region would be subjected to twisting and compression during flight. In deep-diving birds, such as many of the alcids, these processes are elongated, providing greater protection from compression by increased water pressure. Screamers, a small family of South American birds distantly related to ducks and geese, exhibit the opposite condition and completely lack the uncinate process.

Posterior to the thoracic region, a bird's skeleton shows very distinct differences from other vertebrates. A very broad, flattened region known as the **synsacrum** is the result of many bones being fused together. Its components consist of a highly modified pelvis fused with the vertebrae in the lumbar and pelvic region and the first six **caudal** or tail vertebrae. The legs are attached to each side of the synsacrum.

The remaining portion of the vertebral column is the tail, which extends posteriorly from the synsacrum. The bony portion of the tail in birds is very short. It consists of the six short caudal vertebrae that are free of the synsacrum, and the **pygostyle**. The pygostyle is a single, flattened bone made up of the remaining caudal vertebrae which have become fused and very foreshortened. Attached to it are the muscles and connective tissue to which the tail feathers are attached.

The wings attach in the pectoral region of the body, another area where birds show extensive modification when compared to other vertebrates. These modifi-

## Chapter 3 – Avian Skeletal System

cations are essential for the bird to achieve its mode of powered flight. The main bones of the pectoral girdle are the single **sternum** and the paired **coracoids**, **scapulars** and **clavicles**. The right and left coracoid anchor against the anterior portion of the scapula and arise in a nearly vertical column. The scapula and clavicle both articulate against the upper end of the coracoid. The coracoids are the most massive bones of the pectoral region. The shoulder muscles anchor to the scapula, which is not a broad, flat "shoulder blade" such as we have. Instead, each scapula is a flattened, slender bone pointing rearward from the end of the coracoid and lying parallel on each side of the spinal column. In birds, the clavicles have become fused into the familiar V-shaped "wishbone" known as the **furcula**. The upper ends of the furcula articulate against the upper ends of the coracoids. Each clavicle (half of the furcula) is a thin, flexible bone. The coracoid, scapula and clavicle all have a common point of articulation forming a "tripod" upon which the wing is attached. During the down stroke of the wings, the coracoids rotate outward and the ends of the furcula are pulled outward and down. As the pectoralis muscles relax and the wings are drawn upward, the furcula acts like a spring as the clavicles return to their resting position.

### The Wings

The bones of birds' wings follow the same basic plan as the forelimbs of other vertebrates. However, many bones have become fused or lost altogether during the course of evolution.

At the **proximal** (closest to the body) portion of the wing is the **humerus**. This is a strong bone that accounts for approximately 1/3 of the total length of the wing. The main flight muscles attach only to the humerus and no other part of the wing. The forces exerted by these muscles are more easily accepted by a short bone. The actual length of this bone varies considerably between different kinds of birds.

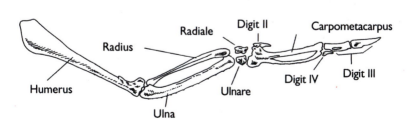

*Figure 3-5. Wing showing skeletal structure only*

## Chapter 3 – Avian Skeletal System

The humerus of an albatross is relatively long while that of a swift is quite short. Attached to the upper surface of the proximal end of the humerus is the tendon from the **supracoracoideus muscle** (sometimes called the **pectoralis minor**) which overlays and originates from the sternum, furcula and ribs. This tendon goes through a channel called the **foramen triosseum**, formed by the scapula and coracoid, and curves up and over to the top of the humerus. When the supracoracoideus contracts, its tendon is tightened and works somewhat like a pulley to raise the wing. This allows the bulk of the muscle mass to be below the wing instead of above the wing on the shoulders where it would more easily upset the bird's center of gravity, crippling its balance and ability to fly. The pectoralis major attaches to the lower surface of the humerus, pulling the wing downward upon contraction.

The elbow is at the **distal** end (most distant from the body) of the humerus. This is the point of articulation with the next major wing bones, the **radius** and **ulna**. The radius and ulna are the bones of the forearm. The radius is a straight, slender bone supporting the anterior portion of the forearm. The ulna is the larger of the two bones and is located more posteriorly in the forearm. It is usually relatively stout and somewhat curved. On the trailing edge of the ulna a single row of tiny projections or bumps are found. These are the points at which the secondary feathers of the wing attach directly into the bone.

The "bend in the wing" of a bird is not the elbow but the wrist. This is located at the distal end of the radius and ulna. The number of **carpels** (wrist bones) in birds has been reduced to only two, the **radiale** and the **ulnare**. The remaining portion of the wing beyond the wrist is the **manus** or "hand." It is here where birds, once again, have fewer bones and many fused bones as compared to other vertebrates. Apart from the radiale and ulnare, the remaining carpels and **metacarpals** of the hand have become fused into a single bone which is called the **carpometacarpus**. During the course of evolution, the first and fifth digits of the hand have become lost. (Many authors consider the fourth and fifth digits to be the absent ones. Whether you consider it to be the first and fifth or fourth and fifth partially depends upon interpretations of developmental evidence. Throughout this book, reference will be made to the second, third and fourth digits, which assumes that the first and fifth are absent.) The second digit (sometimes called the thumb by those who consider it to be first digit) is formed by two fused **phalanges** (finger bones). It joins the hand at the junction of the radiale and carpometacarpus. These bones, and the feathers associated with them, form the **alula**, or second wing (sometimes referred to as the "bastard wing"). (See Chapter 2 on flight for a discussion of the alula's function.) The third digit is at the outer end of the carpometacarpus. Two of its three phalanges are fused together. The fourth digit is a single phalanx (singular of phalanges) which is attached at the joint of the carpometacarpus and the phalanges of digit III. The structure of a bird's hand contrasts sharply with that of a bat, which has long, well-developed phalanges with a webbing of skin forming the wing.

## Leg Bones

A common misconception is that a bird's knee points rearward when it bends, opposite that of the human knee. If we examine the leg bones of a bird, it is clear that the knee works in the same manner as the human knee and it is the ankle that actually points to the rear. In life, muscles and feathers cover the upper leg and knee joint, obscuring it from our vision. In birds, as in other limbed vertebrates, the **femur**, or upper leg bone, is attached to the pelvic girdle. Below the knee, the leg shows some modification of a "typical" vertebrate leg plan. The **tibia** and **fibula** are fused into a single, rather long bone known as the **tibiotarsus**. At the distal end of the tibiotarsus is the ankle joint and the beginning of the foot, which has become highly modified in birds. The legs and feet of birds are covered with scaly plates which give strength and provide protection.

Terrestrial vertebrates (except snakes) walk with only their toes hitting the ground (like cats) or with the full foot (heel to toe) striking the ground (as in humans). Those animals which walk on their toes are called **digitigrade** while those that walk with their heels striking the ground are called **plantigrade**. Birds use their toes and thus are digitigrade. On occasion an individual bird, especially a nestling, will rest with its heel upon the ground, but most often the posterior portion of a bird's foot is elevated above the ground. The bones in this part of the foot are fused together into one large, extended bone called the **tarsometatarsus**. It may also be referred to as simply the **tarsus**.

## The Foot

Most birds have four toes. The first toe, analogous to our big toe, is called the **hallux**. It consists of two phalanges and generally points rearward in most birds. The second or innermost toe has three phalanges, the middle toe has four phalanges and the outermost toe has five phalanges. The fifth digit is absent in all birds.

The most common foot pattern in birds is called **anisodactyl**. This pattern has digits II, III and IV pointing forward and the hallux (digit I) pointing rearward. Birds in the Order Passeriformes, as

## Chapter 3 – Avian Skeletal System

well as many other birds, have this foot structure. While this is the typical pattern, numerous other patterns of foot structure are found in birds.

The second most common foot pattern is **zygodactyl**. In the zygodactyl foot, digits II and III are forward and digits I and IV are rearward. In some birds (such as Osprey) with this arrangement, the fourth toe can be rotated forward or rearward. Zygodactyl feet are found in Osprey, owls, cuckoos, most woodpeckers and parrots, mousebirds and some swifts. Trogons have a superficially similar foot called **heterodactyl**. It looks much like a zygodactyl foot with two toes forward and two toes rearward, but in the heterodactyl foot digits III and IV point forward and I and II point rearward. This type of foot is found only in the Family Trogonidae.

The **syndactyl** foot, found in kingfishers and hornbills, has digits II, III, and IV forward and the hallux rearward as in the anisodactyl foot, but digits II and III are partially fused along much of their length. In the **pamprodactyl** foot, digits II and III are forward and I and IV are free to pivot either forward or rearward. This is the foot pattern found on most swifts. Swifts cannot perch on a branch like songbirds. Instead, they cling to the sides of crevices on cliffs or inside hollow trees. This is usually better accomplished by having all four toes forward to act as hooks against the surface.

**Raptorial** feet are similar in pattern to anisodactyl feet with digits II, III and IV forward and I rearward pointing, but the toes are very heavy and strong. They are also armed with large sharp claws or talons for catching and holding prey. In many of these birds, especially falcons, the center or third toe is greatly elongated. Kites, eagles, hawks and falcons have raptorial feet. Owls may be considered birds of prey, but have zygodactyl feet.

Many birds have webbing between toes or along the margins of toes. There is considerable variety of webbing patterns between different groups of birds. The most common webbing pattern is that found in ducks, geese, gulls and many other aquatic birds and is called **palmate**. A palmate foot has digits II, III and IV forward and I to the rear. There is full webbing between all of the forward toes. The hallux is frequently short. When there is only a partial webbing between the toes (that is, it does not reach to the ends of the toes) it is considered to be a **semipalmate** foot pattern. This type of foot structure is found in several sandpipers and plovers and many members of the Order Galliformes. In some aquatic birds, especially of the Order Pelicaniformes, webbing occurs between all four toes and

## Chapter 3 – Avian Skeletal System

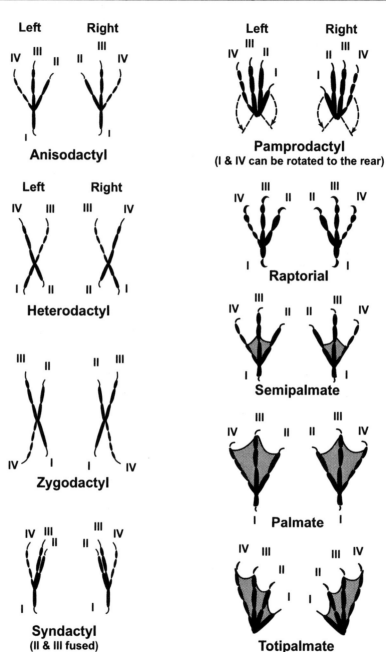

Figure 3-6. *Toe and foot structures of different types of birds*

all digits are directed forward. This foot pattern is known as **totipalmate**.

In some kinds of aquatic birds there is a web of skin extending out from each side of the toes but not a connecting web between the toes. This is called a **lobate** foot. Loons, grebes and coots have this type of foot pattern. When the bird pushes its foot back against the water while swimming, the toes and their webbing expand, providing a large surface area to work against the water. As the foot is brought forward, the lobes fold back against the toe to decrease resistance.

# Chapter 4
# Circulation and Respiration

Flight requires a large expenditure of energy and results in a considerable buildup of heat. An efficient circulatory system is needed to supply a bird's rapidly working muscles and demands for glucose and oxygen. Likewise, a very efficient respiratory system is necessary to supply the volume of oxygen needed, to remove excess carbon dioxide, and to help dissipate excess heat.

## Circulatory System

Birds share some characteristics with their reptilian ancestors, but the structure of their circulatory system is more like that of a mammal. Reptiles are often called "cold-blooded" while birds and mammals are called "warm-blooded," but these are inaccurate terms. The temperature of the blood can vary (especially in reptiles) with the ambient temperature of the surrounding air, and with any vertebrate the temperature of the blood may vary in different parts of the body – blood in the foot is usually cooler than blood in the interior of the body, for example. Better terminology would refer to reptiles as **poikilotherms** and birds and mammals as **homeotherms**. A poikilotherm cannot regulate its metabolism to maintain a relative constant body temperature. Homeotherms do maintain a constant body temperature, regardless of the temperature of the surrounding environment (air or water).

Birds also share with mammals the characteristic of having a four-chambered heart. Reptiles have a three-chambered heart with two atria and one ventricle. This single ventricle has a septum, or internal wall dividing it but it is incomplete. This results in freshly oxygenated blood becoming partially mixed with the oxygen-poor blood returning from the veins of the body. This mixed blood carries less oxygen to the body's tissues. Such a system does not supply enough oxygen to meet the metabolic requirements of a homeotherm. The four-chambered heart, with two atria and two ventricles, prevents oxygenated blood from the lungs from mixing with unoxygenated blood from the veins.

### The Pathway of the Blood

The **atria** (**atrium**, singular form) are the upper chambers of the heart. They are small and thin-walled in comparison to the **ventricles**. The ventricles are the large, very muscular portions of the heart and account for the majority of the heart's mass. The ventricles are stronger and more muscular because they need to supply sufficient force to pump the blood away from the heart and out through the body. Each atrium needs to only pump the blood past the valve that separates it from the adjoining ventricle directly below. Each heartbeat has two steps: First, the atria contract, then the stronger contraction of the ventricles follows. Thick-walled, flexible arteries carry blood away from the heart and the thin-walled veins carry blood returning to the heart from the body.

In birds, two **precava** veins carrying blood from the anterior region of the bird join into a single **sinus venosis** as they enter the right atrium. (Mammals have a single **superior vena cava** instead). Blood returning from the posterior regions of a bird's body enters the right atrium through the **inferior vena cava**. At this point, the blood is low in oxygen but high in carbon dioxide ($CO_2$). As the right atrium contracts, blood is pumped into the right ventricle immediately below it. When the right ventricle contracts, blood moves away from the heart via the **pulmonary trunk** which immediately splits into the right and left **pulmonary arteries**, transporting blood to the right and left lung respectively. In the lungs, $CO_2$ leaves the blood, while oxygen is taken up by the **hemoglobin** molecules of the blood. This oxygen-rich blood returns to the heart and enters the left atrium. Contraction of the left atrium moves blood into the left ventricle. When the left ventricle contracts, this oxygenated blood leaves the heart through the **aorta** and moves throughout the arterial system of the body. It eventually reaches the **capillaries** throughout the body's tissues where gases are exchanged ($O_2$ from blood to tissues, $CO_2$ from tissues to blood), and the blood then returns to the heart via the veins.

A series of valves, flaps of tissue which open or close, help prevent blood from flowing backward through the system. The **tricuspid valve** separates the right atrium from the right ventricle. The three flaps of tissue which form this valve are forced apart as the atrium contracts, pushing blood through the valve into the ventricle. Contraction of the right ventricle opens the **pulmonic valve,** allowing blood into the pulmonary arteries. The **mitral valve** is located between the left atrium and left ventricle. The valves in the heart close with enough force to make an audible sound. It is this sound, not the muscular contraction, which makes the "heartbeat" that you can hear. (The same is true for humans. When the doctor listens to your heart, it is these valves snapping shut that she/he listens to.)

## Heart Rates & Blood Pressure

The left ventricle is the most massive portion of the heart, and in birds, the thickness of the ventricle wall is up to three times that of the right ventricle. The blood pressure generated by contraction of the left ventricle is five to

## Chapter 4 – Circulation and Respiration

ten times greater than that of the right ventricle. It is important that the right ventricle be less powerful, as it needs to only be strong enough to pump blood through the lungs, not through the rest of the body as does the left ventricle. A smaller sized ventricle produces a smaller contraction with a much lower blood pressure. Pressure as great as that produced by the left ventricle could damage

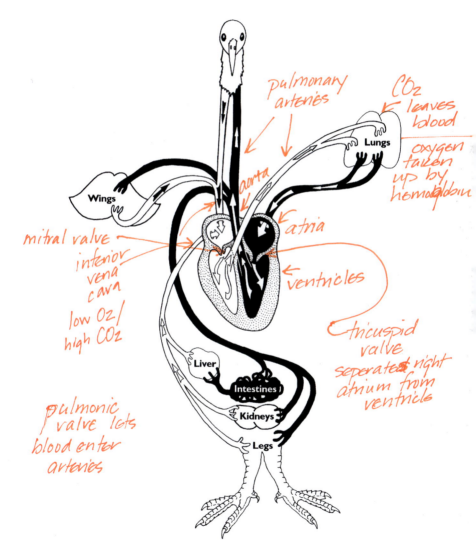

Figure 4-1. Pathway of blood flowing through a bird's body. Dark areas indicate oxygenated blood, white areas in the veins and vessels indicate unoxygenated blood.

the lungs by rupturing the many tiny vessels and capillaries in the lungs. The lower pressure from the right ventricle prevents such damage. Lower pressure does not mean lower volume. The quantity of blood pumped by each of the heart's chambers must be identical to keep a constant flow of blood throughout the body.

When a bird and mammal of equivalent body mass are compared, the bird will be found to have the larger heart. The bird's heart also has more muscle fibers and is able to sustain a higher level of activity, such as that required for flight. The larger heart of a bird pumps more blood per heartbeat (**stroke volume**) than does a mammal, and birds at rest typically have a slower heartbeat than a similar-sized mammal. Activities such as flight quickly elevate the heart rate. Small birds have a more dramatic increase in heart rate with exercise than do large birds.

| Resting Heart Rates | |
|---|---|
| Species | Heart rate |
| Domestic turkey | 93 |
| Mallard | 113 |
| Herring Gull | 218 |
| Rock Dove (pigeon) | 166 |
| American Crow | 342 |
| American Robin | 328 |
| House Sparrow | 350 |
| House Wren | 480 |
| Ruby-throated Hummingbird | 615 |

*Figure 4-2. Comparison of resting heart rates (expressed as beats per minute) for several North American birds*

A large gull may go from 150-200 beats per minute while at rest to over 600 beats/minute during flapping flight. Hummingbirds may exceed over 1200 heartbeats per minute while flying!

In birds, the relatively larger ventricle and its stronger contraction result in a higher blood pressure than in mammals. Male birds usually have a slightly higher blood pressure than do females of the same species. The European Starling, *Sturnus vulgaris*, has a blood pressure of 180 mm Hg (millimeters of mercury). A human with a blood pressure of 150 mm Hg is considered to have high blood pressure. Domestic turkeys are reported to have blood pressures as high as 300-400 mm Hg. This is very high, even for birds, and can lead to death by rupture of the aorta. Even small songbirds, under extreme stress, can suddenly die from a ruptured aorta. The arteries of birds tend to be somewhat more thin-walled and less flexible than the arteries of comparable sized mammals. As a result, birds are readily susceptible to atherosclerosis, an additional problem for the turkey farmer trying to raise large, fat turkeys for market.

## Avian Blood

The structure of a bird's circulatory system is rather mammal-like, but the blood is more similar to the blood of reptiles than it is to the blood of mammals. In mammals, mature **erythrocytes**

## Chapter 4 – Circulation and Respiration

(red blood cells) are cells that are biconcave in shape and have lost their nuclei. They are densely packed with **hemoglobin**. Hemoglobin is the molecule responsible for transporting oxygen ($O_2$). In many invertebrates, hemoglobin is dissolved in the plasma of the blood, but most vertebrates have hemoglobin packed in the erythrocytes rather than in the plasma. Like those of mammals, bird erythrocytes are filled with hemoglobin, but unlike mammals, the cells remain nucleated and are biconvex in shape. The hemoglobin content in birds' blood is slightly less concentrated than that in mammals, and has a slightly different structure that allows it to be more efficient at carrying oxygen. As in mammals, carbon dioxide is dissolved in and carried by the blood plasma.

As carbon dioxide dissolves in the blood, carbonic acid is formed and the pH of the blood drops (a pH value of 7.0 is considered neutral. Values from 1 - 6.9 are acid and values from 7.1 - 14 are alkaline). The pH of the blood is normally 7.3 - 7.4. In humans and other mammals, sensory cells in the circulatory system detect this drop in pH and tell the brain to increase the rate of breathing. Thus, as we exercise, we begin to breathe harder and faster. The increased ventilation rate should expel $CO_2$ more quickly and bring more $O_2$ into the system. Rapid breathing, or hyperventilation, can create an opposite problem.

As large amounts of $CO_2$ are eliminated, the pH of the blood increases (becomes more alkaline). This increase causes blood vessels to constrict reducing the flow of blood to the brain and decreasing the amount of oxygen available to the brain. If continued, the blood flow may decrease by 50% or more, resulting in a loss of consciousness or, with severe decrease, death. This is even more of a problem at higher altitudes where the oxygen available in the atmosphere is at a lower concentration. Many birds fly at high altitudes, especially during migration, and are breathing rapidly because of the lower oxygen level and the exertion of flapping their wings. This can increase the alkalinity of the blood to as much as pH 8, a condition that would be fatal to any mammal. It is not known why birds can survive this, but it is fortunate for them that they can.

## Respiratory System

Flight requires an efficient respiratory system, able to deliver large volumes of oxygen to working muscles. Birds have the most efficient and structurally most unique respiratory system of all vertebrates. In addition to delivering the high volumes of oxygen demanded by sustained flapping flight, the respiratory system helps to disperse excess heat built up during vigorous exercise.

## Chapter 4 – Circulation and Respiration

### Lungs

In mammalian lungs, air is drawn in through the nose or mouth and passes through the **trachea**, which divides into the **bronchi** (one for each lung) that carry the air into the lung itself. As the bronchi enter the lungs, they divide over and over into smaller and smaller units until finally they end in a multitude of microscopic air chambers called **alveoli**. Clusters of these alveoli fill the space of the lung. But these form a dead-end to air passageways. As we exhale, the air just brought into the lung is reversed in direction and passes out of the lungs through the same vessels which brought it into the lungs. There is always some residual air left in the lungs, thus the system is not 100% efficient. The efficiency of this system works well enough for mammals but falls short of the needs required for the flapping flight of birds.

Birds do not have the two-way flow of air that is found in mammals. Instead, birds have air flowing in only one direction and passing through the lungs rather than in and out. There are no dead-end chambers and no residual air to be left behind. Fresh air is passing through the lungs nearly 100% of the time. This flow is assisted by a series of **air sacs** (see below) which are unlike anything found in any other kind of vertebrate.

Avian lungs are small, about half the volume of a similar-sized mammal, but they are composed of much more compact, dense and highly vascularized tissue. They lie within the rib cage alongside the ribs but do not fill most of the space within the chest cavity as do mammal lungs. Internally, the lung has many highly branched vessels through which air passes.

In mammals, a muscular diaphragm separates the thoracic and abdominal cavities. Regular contractions of this diaphragm cause the lungs to inflate and partially deflate, creating a tidal flow of air in and out of the lungs. Birds have no diaphragm and generate airflow through the lungs by expanding and contracting the rib cage. A bird inhales by lowering the sternum, which expands the chest cavity which, in turn, enlarges the air sacs and draws air in through the lungs. To exhale, the bird compresses the sternum and rib cage, compressing the air sacs. Until recently, it was widely believed that movement of air into and out of the body coincided with wing beats during flapping flight. Flexion of the furcula during flight enhances the movements of the sternum but the rates of breathing and flapping are maintained independently.

### The Route of Air

With the tidal flow of air in human lungs, each exhalation expels air brought into the lungs by the previous inhalation. In

## Chapter 4 – Circulation and Respiration

birds, two breathing cycles are required to complete the passage of air through the respiratory system. Also, unique to birds, air is continuously moved *through* the lungs in a one-way flow. During the first inhalation, air is taken in through the nostrils at the base of the bill (and sometimes, through the mouth) and drawn down the **trachea**. The trachea splits into two **bronchi**, one entering each lung. Each of these **primary bron-** chi, now called a **mesobronchus**, passes entirely through the lung, emptying into the **posterior air sacs**. At the first exhalation, the abdomen contracts, compressing the posterior air sacs and forcing air to return to the lungs, but through a different set of vessels, the secondary bronchi. These secondary bronchi divide again into **tertiary bronchi**, called **parabronchi**, each less than 0.5 mm in diameter. Branching from

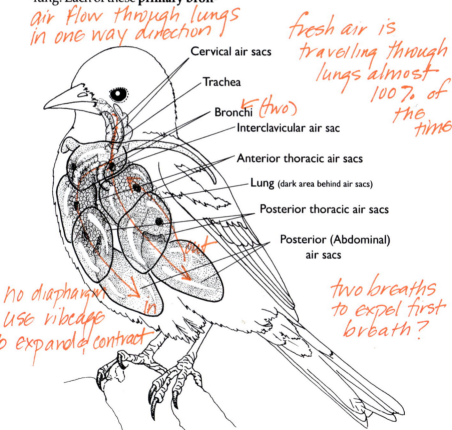

Figure 4-3. Birds have two lungs, and typically nine air sacs associated with the respiratory system, as shown here.

these are even smaller **air-capillaries** which are surrounded by many blood-capillaries. It is here where gas exchange ($O_2$ from lung to blood; $CO_2$ from blood to lung) occurs. This network of air-capillaries and adjoining blood-capillaries form the bulk of the lung tissue. Birds such as ducks, which are strong, long-distance fliers, may have up to 1800 parabronchi in each lung. Chickens and other weak fliers may have as few as 400 parabronchi per lung. Unlike the alveoli of mammalian lungs, the parabronchi are not dead-end channels but one-way tubes which come together again. During the second inhalation, the air is continued through the lung to the **anterior air sacs**. On the second exhalation, the anterior air sacs are compressed. This air is returned to the nostrils where it is now expelled, completing this two-phase cycle.

Breathing rate is a function of an animal's size and level of activity. Large birds have a slower breathing rate than do small birds. A turkey may only breathe seven times per minute but a small hummingbird may breathe over 140 times per minute. Muscles in flight have a high demand for oxygen and breathing may increase 12 to 20 times over the rate at rest.

In many birds, some form of protection, such as a dermal flap (**operculum**) is provided to the nostrils. In diving birds, this operculum covers the nostrils and prevents water from entering. In woodpeckers, the nostrils are filled with many tiny bristles which prevent small pieces of wood from entering the nostrils while the bird is excavating a hole.

## Air Sacs

The air sacs of birds are unlike any structures found in the respiratory system of any other vertebrate, either in form or function. The air sacs are an important part of the respiratory system and are directly connected to the primary and secondary bronchi. They can occupy as much as 15% of the volume of a bird's abdomen and thorax, yet they are very thin-walled, only two to three cell layers thick. In many birds, some of these sacs extend into the hollow center of the large bones. The blood circulation surrounding these sacs is small and indicates that these sacs play little or no role in $O_2$ - $CO_2$ exchange. Instead, air flowing over the extensive moist surfaces provided by these sacs allows for evaporative cooling, an important function which prevents a bird from overheating while flying. Expansion and contraction of the body walls compress and expand these air sacs, making a very efficient flow of air through the bronchi and lungs possible. These sacs also provide a degree of cushioning and protection to the internal organs of a bird. The single large sac in the front of the breast is especially enlarged in diving birds such as Brown

*air sacs thin walled - 2-3 cells thick. in many birds these sac go into hollow center of large bones - provides evaportive cooling.*

Pelicans and probably provides a cushion as the bird hits the water.

Most birds have nine air sacs. The **cervical air sacs** are paired on either side of the neck. Some males overinflate these sacs during breeding displays. Male Greater Sage-Grouse, *Centrocercus urophasianus*, and other kinds of grouse inflate the cervical air sacs, expanding the overlaying skin into visible bulges while displaying to attract females on their breeding grounds (called a lek). The bright red throat pouch of the displaying male Magnificent Frigatebird, *Fregata magnificens*, is caused by hyperinflation of the cervical sacs.

Across the front of the thorax is the single **interclavicular air sac**. Branches of this sac extend out into the **pneumaticized** (hollow) portions of the sternum and major wing bones. A broken humerus, with the broken portion exposed to the air, would allow air into the interclavicular air sac and allow a bird to breathe even if the trachea were completely blocked, preventing air from passing through the nostrils! A portion of this air sac is also found in the syrinx, the vocal organ of birds. Changes in pressure within this air sac affect membranes in the syrinx and are essential for a bird to vocalize.

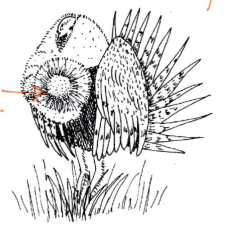

*Figure 4-4. Male Greater Sage-Grouse inflate frontally-directed cervical air sacs which cause a popping sound as they are inflated and deflated during breeding displays.*

Puncturing this air sac would render a bird voiceless.

Behind the interclavicular air sac are two pairs of air sacs: the **anterior thoracic air sacs** and the **posterior thoracic air sacs**. At the rear of the body cavity are the two largest air sacs, the paired **abdominal air sacs**.

Nine is the typical number of air sacs, but a few birds have less and some have more. Many shorebirds have 12 air sacs, but Loons have only seven. African weavers have only six sacs, the lowest number found in birds.

## Chapter 4 – Circulation and Respiration

*rhamphotheca: horny sheath covering mandibles*

# Chapter 5
# Feeding Adaptations & Food Gathering

Birds are high-energy creatures and as such, have high demands for food. The quest for food occupies a major portion of a bird's life. Birds have many unique adaptations which aid in the gathering of food.

## Bill Structure

The bill or beak of a bird is a unique structure in the animal world. It does not contain teeth as do mammalian jaws, but must serve some of the same functions. A common misconception is that the bill or beak is a rigid structure of nonliving tissue much like our fingernails, but this is not the case. In most birds, the upper **mandible (maxilla)** can be bent upward by flexing the **nasofrontal hinge**, a flexible sheet of thin bone which attaches the upper portion of the bill to the skull. Some birds can even flex the tip of the maxilla itself. A woodcock, probing deep into the mud, grasps its prey by

**Whimbrel**
Prober

opening only the tip of the bill. The lower mandible is attached to the skull and operated by large, powerful muscles. These muscles are especially powerful in birds such as grosbeaks, which crack large seeds. Both mandibles are covered by a thin horny sheath (**rhamphotheca**). Numerous nerve endings, allowing taste and/ or touch, occur along the edge of the bill and in the palate of many birds. Sanderlings, probing below the surface of the tidal sands, actually taste the sand for the presence of the invertebrates upon which they prey. To fully appreciate the sensitivity of birds' beaks, consider that the margins and the palate of a Mallard contain more touch receptors than the most sensitive area of human skin.

The shape of the bill determines much of what a bird's diet will be, such as the hooked bill of a raptor for tearing flesh, or the stout, conical bill of a seedeater.

**Evening Grosbeak**
Seed Cracker

## Chapter 5 – Feeding Adaptations and Food Gathering

### Name of Birds' Bill Shapes

**LONG:**
The bill is noticeably longer than the head. (Great Blue Heron)

**SHORT:**
The bill is shorter than the head. (American Goldfinch)

**COMPRESSED:** *higher than wide*
The bill is higher than it is wide for much of its length. (Tufted Puffin)

**DEPRESSED:** *wider than high*
The bill is wider than it is high. (Mallard)

**RECURVED:** *upward*
The mandibles turn upward at the tip. (American Avocet)

**DECURVED:** *downward*
The mandibles turn downward at the tip. (Brown Creeper)

**TERETE:** *Circular in cross section*
The bill is more nearly circular is cross-section along most of its length. (Rufous Hummingbird)

**BENT:**
The bill is curved (usually downward) at a strong angle along the middle of its length. (Flamingo)

**HOOKED:**
The upper mandible is longer and sharply bent over the lower mandible at its tip. (Bald Eagle)

**SPATULATE:** *spoon shaped*
The bill is wide and spoon-shaped near its tip. (Northern Shoveler)

**CROSSED:**
The tips of each mandible cross over one another. (Red Crossbill)

*Figure 5-1. Bill Shapes*

---

Many terms are used to describe bill shape. Some of these terms are defined in the chart above. A variety of probing shapes are found in shorebirds, where each shape is associated with feeding strategies. These shapes range from the very long beaks (Long-billed Curlew) found on birds that probe deeply into the ground for the deeper-dwelling invertebrates, to short beaks (Semipalmated Plover) of birds which probe at or just below the surface. The short beaks of turnstones are slightly flattened to allow these birds to live up to their names. Their beaks are used to pry under and overturn small stones to expose invertebrates which they readily consume. Sometimes, they even work in groups of two or more to flip over a stone too large for one bird to overturn alone. Many seed-eating birds are able to quickly open seeds with their well-adapted bills. In many of these birds, the upper mandible has a groove along its inner edge and the lower mandible has a sharp ridge (see Fig. 5-2). A seed is held in the groove while the lower mandible's edge cuts into the seed coat, usually at its margin or weakest point. The lower mandible is forced up and

## Chapter 5 – Feeding Adaptations and Food Gathering

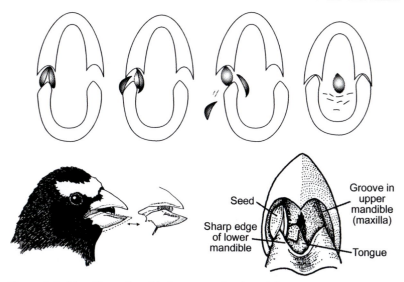

Figure 5-2. Top: Shows how birds open seeds; above left, shows how birds roll their beaks forward and backward to position seeds; above right, shows structure of beak with seed positioning indicated.

forward, cracking and removing the seed coat while the tongue extracts the seed. Some finches are able to exert over 100 pounds of force to open a hard-walled seed. Figure 5-3, 5-4 and 5-5 show a variety of birds and the various types of beaks.

## Digestive System

In addition to the specialized adaptations of the bill, the avian digestive system is also uniquely adapted for a bird's high-energy lifestyle. Birds have a high metabolism and need to eat frequently. A flying bird must keep its weight to a minimum and cannot afford to have heavy food in its stomach for long periods of time. Birds usually eat small quantities but do so very often throughout the day. A small bird which spends long periods of time foraging may put itself at risk from predators. It must balance nutritional gains against time spent food-gathering. Large insects often have more nutritional value for birds than do some small insects, but it may be safer for a bird to quickly gather several small insects than to spend the time looking for fewer large, but less abundant insects.

The route through the avian digestive system begins with the beak and oral cavity. Birds have no teeth so very little processing of food takes place in the mouth. Seed coats are cracked and removed and some insects may be partially crushed before being

## Chapter 5 – Feeding Adaptations and Food Gathering

**Pileated Woodpecker**
Chisels into wood for insects

breeding season, as much as 50 times. Some North American swifts have salivary glands that enlarge about 12 times during the breeding season.

The tongue lies within the oral cavity, but it is not a muscular organ as it is in mammals. It assists with gathering and swallowing food and contains many nerve endings for touch and probably taste. The shapes of bird tongues vary greatly. Most birds have small projections on the tongue which help force food down the throat. The specialized tongues of nectar-feeders are fringed and "feathery" on the end to help take up the nectar. Filter-feeding birds, such as the Northern Shoveler, have very elaborate tongues designed to serve as fine sieves in the water.

swallowed. Bird's saliva acts as a lubricant; some have saliva that is further specialized. Woodpeckers have a sticky saliva which helps to hold insects. The saliva of some jays is also sticky and is used to form a ball of food (**bolus**) to be stored in a crevice or elsewhere for later use. Some swallows, swifts and swiftlets have saliva which acts as a type of cement and is used to anchor the nest or eggs to a substrate. The Edible-nest Swiftlet, *Collocalia*, found in the East Indies, has a nest which is a small shelf made entirely of dried saliva. It is this dried saliva nest that is the basic ingredient of Bird's-nest soup. The glands that produce the saliva in these birds greatly enlarge during the

Once past the bill and oral cavity, the food enters the esophagus. In humans, the esophagus is primarily a passageway to the stomach, but in birds it may have many more functions. In some birds (Ostrich, bustards), it can be inflated for display or used as a sound resonator. Pigeons and doves secrete a milky substance ("pigeons milk") to feed their chicks. Although not produced by mammary glands, as in mammals, this milk is similar in composition to mammalian milk with similar proportions of fats and proteins.

In most birds, a portion of the esophagus has developed into a structure known as the **crop**. The crop helps to regulate the flow

*diverticulum: saclike projection in esophagus.*
*owls don't have crop*

## Chapter 5 – Feeding Adaptations and Food Gathering

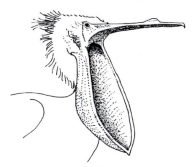

**White Pelican**
Fish Trap

of food into the stomach and, in some species, it may store food. A pelican does not store food in its bill, as commonly believed, but rather it carries fish in its crop. The bird later swallows these fish or perhaps regurgitates a portion to feed its young. The crop may be a simple expansion of the esophagus or an enlarged, saclike projection known as the **diverticulum**. The diverticulum of crossbills and redpolls is used to store seeds until ready to be digested at a later time, sometimes as much as a day later. The crop is usually much more developed in seed or grain-eating birds. However, in fish-eating birds, like cormorants, the crop is typically a simple expansion of the esophagus and in owls, a crop is completely lacking.

Few birds are true herbivores, as leaves and coarse plant matter take a long time to break up and digest. Mammalian ruminants, like cattle, chew and regurgitate food for several hours. Flying birds simply cannot spend the time and energy necessary for this process. Two interesting exceptions are the Hoatzin of South America and the Kakapo, a large, nocturnal parrot of New Zealand. In both of these birds, the crop has become very well-developed and acts as a large, muscular stomach, which, in fact, is many times larger than the actual stomach.

*proventriculus: anterior stomach*

*gizzard: posterior stomach*

**Great Blue Heron**
Medium-sized Fish Grabber

Beyond the esophagus and crop, the next portion of the digestive system is the stomach. In birds, the stomach is a two-chambered structure. The forward or anterior portion of the stomach (**proventriculus**) is a glandular organ and the latter, or posterior portion (**gizzard**) is thick-walled and muscular. The most developed proventriculi are found in fish-eating birds and raptors. It is in the proventriculus where secretions of highly acidic (pH 0.2 - 1.2) peptic enzymes and hydrochloric acid are made. These aid the bird in its ability to digest bones and flesh. A shrike can digest a mouse in less than three hours and a Lammergeier or Bearded Vulture, *Gypaetus barbatus*, can digest an entire cow vertebra in two days!

**Birds! From the Inside Out** 47

## Chapter 5 – Feeding Adaptations and Food Gathering

Petrels store oils, by-products of digestion, in the proventriculus. These oils are regurgitated to feed the young and as a defense against predators (or ornithologists who approach too closely, upon whom the Petrels might spit).

The **gizzard**, or muscular portion of the stomach, grinds the food and is especially strong in grain-eating birds such as turkeys. These birds usually have small stones or grit in the gizzard to assist with grinding. It is interesting to note that these birds are able to retain the grit while allowing food to pass through the stomach, thus eliminating the need to constantly search for new grit. In addition to this grit, the inner surface of the gizzard is often hard and keratinized. It contains many folds, providing numerous hard grinding surfaces. Early experiments with turkeys, *Meleagris gallopavo*, demonstrate some amazing results. Réaumur, in 1752, found that a tube of sheet iron was flattened and partially rolled after only 24 hours in a turkey's stomach. This was the same steel that could not be dented until subjected to 80 pounds force! He also found that the turkey could pulverize 24 English walnuts in the shell in just four hours. Even more incredible is the work of Spallanzani in 1783. He found that 12 steel needles were completely ground in the stomach of a turkey in 36 hours and that 16 surgical lancets were ground in 16 hours! In contrast, the gizzards of raptors, fish-eating and fruit-eating birds are rather thin-walled, as these birds depend more upon enzymatic activity than mechanical breakdown of food for complete digestion. The gizzard may change in structure as the diet changes from season to season in some birds. A bird, such as a Black-capped Chickadee, that eats insects in the summer may have a small, thin-walled gizzard at that time. This same bird will have a much larger and thicker gizzard in the winter when a primary food source at that time is seeds.

At the end of the digestive tract are the intestines and **cloaca**. In some birds, there are a pair of simple sacs, called **ceca**, which divert from the side of the large intestine near its posterior end. These ceca contain bacteria which probably assist in digestion of less-easily digested materials. Large ceca are present in ground-dwelling birds but nearly absent in most flying birds where long time periods for digestion and watery stools would add unacceptable weight.

## Chapter 5 – Feeding Adaptations and Food Gathering

**Black Turnstone**
Flattened tip for prying under and flipping stones

**Hooded Merganser**
Serrated edge for grabbing and holding small fish

**Red Crossbill**
Crossed tips used to pry open conifer cones

**Common Poorwill**
Large mouth, edged with bristles for catching flying insects

**Dunlin**
Probes below surface for food

**Wood Duck**
Plant-Eater: acorns, nuts, aquatic plants

*Figure 5-3. Structural Adaptations in Bird Beaks*

**Birds! From the Inside Out**

## Chapter 5 – Feeding Adaptations and Food Gathering

**Black Oystercatcher**
Quickly opens mussels by stabbing through the hinge of the shell

**Flamingo**
Sifts mud with elaborate beak and tongue

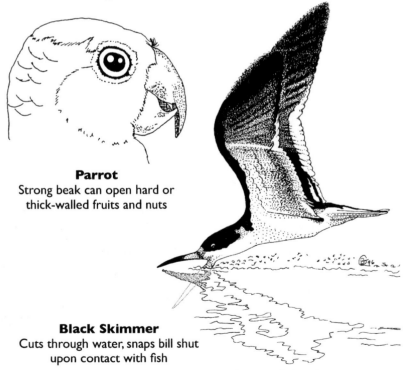

**Parrot**
Strong beak can open hard or thick-walled fruits and nuts

**Black Skimmer**
Cuts through water, snaps bill shut upon contact with fish

*Figure 5-4. Structural Adaptations in Bird Beaks*

## Chapter 5 – Feeding Adaptations and Food Gathering

Figure 5-5. *Structural Adaptations in Bird Beaks*

# Chapter 5 – Feeding Adaptations and Food Gathering

# Chapter 6
# Feeding Behaviors

In American slang, to be "bird-brained" is to be considered slow, simple-minded or stupid. Yet, if we examine the truth about birds we might change our opinion and perhaps even feel complimented by the use of this term. As we will see, birds are quick, highly adaptable, and often show signs of high intelligence.

Behavior and intelligence are large and complex issues of study. In birds, it may often be difficult to separate innate or instinctive behavior from learned and intelligent behavior. These various behaviors often work hand-in-hand (wing-in-wing?) to assist the bird in such activities as food-gathering, nest-building, courtship and mating, and defense.

Birds often have to be very resourceful when in search of food. Collecting sunflower seeds from a feeder may be a very easy task but many more skills are needed when searching out live prey which may be scarce or well hidden in its environment. Some birds have even taken to tool-using in order to secure food. Perhaps the best known example of this is the Woodpecker Finch on the Galápagos Islands. There are no woodpeckers on the Galápagos and this finch has evolved to assume a lifestyle similar to woodpeckers. However, its beak is not as long as a true woodpecker and it sometimes has difficulty extracting insect larvae burrowed deep into the wood. In this case, the resourceful bird has learned to use a small stick or cactus spine which it holds with its beak and uses to pry the insect out to where it can be easily eaten. The Mangrove Finch, another of "Darwin's Finches," lives on a different island where the Woodpecker Finch is not found.

*uses stick to get to bugs when beak can't reach.*

*Figure 6-1. A Green Heron drops "bait" onto the water surface and waits for a fish to come investigate.*

## Chapter 6 – Feeding Behaviors

Like the Woodpecker Finch, it has similar tool-using techniques, which it uses to pry the insect out to where it can be easily eaten.

The Egyptian Vulture, another tool-using bird, is found in Africa. As part of its diet, this bird eats the contents of the eggs of other birds after it cracks open the shells. Ostrich eggs are a favored food but the shell of the ostrich egg is too thick for this small vulture to crack with its beak. The vulture simply finds a rock which it throws at the egg repeatedly until the shell is cracked. It then pries apart the shell and eats the egg inside.

*uses rocks to open ostrich eggs.*

Figure 6-2. Great Tit – In Europe, this species learned to shred the milk tops, in order to gain access to the cream in the milk bottle.

One need not travel to exotic locations to find examples of tool-using or resourceful feeding behavior. On several occasions, Green Herons have been observed baiting fish. The heron places small food objects such as small pieces of bread carefully upon the water. It makes no attempt to eat the food and in fact will actively drive away other birds who do try to eat it. It simply waits. The food placed upon the water attracts the attention of small fish which are then quickly eaten by the patient heron.

The ability of birds to learn and adapt to new situations has been demonstrated many times. Since the 1920s in Great Britain, tits (similar to our chickadees) have been observed removing the tops from milk bottles to drink the cream. The milk bottles, delivered to people's doorsteps, were capped with cardboard or foil. The birds removed, punctured or simply tore away the caps in strips until the cream was able to be reached. Tits shred paper-like bark to obtain nesting materials, so this practice may be a natural extension of that behavior. The behavior was first observed in Great Tits and Blue Tits but other birds apparently watched the tits and learned to open milk bottles as well. This practice has now been reported for over 11 species of birds. With the use of modern milk containers which are more difficult to open and fewer door-to-door deliveries, this behavior is now much less common.

Some birds display a remarkable gift of memory. Chickadees may store thousands of seeds which they return to and feed upon over the winter. Perhaps the best ability to remember is that found in the nutcrackers. In one season, a Clark's Nutcracker can store as many as 30,000 seeds which it will retrieve up to six months later.

Figure 6-3. Clark's Nutcracker visits one of its several thousand food caches.

These seeds are stored in caches of two to three seeds each, which means that a single bird must accurately remember the location of 10,000-15,000 caches each year! This is an astounding ability, especially when you consider that over 70% of the seeds are recovered each year. Were those seeds not recovered simply forgotten? Possibly, but some seeds will have been eaten by other species and some simply may not have been needed. In any case, remembering that many locations accurately is well beyond what most humans could do.

It is sometimes difficult to know what is instinctive and what is learned behavior. A phalarope spins around in the water stirring up and feeding upon prey organisms; a pheasant pecks at the ground to obtain seeds or insects to eat. These seemingly simple behaviors may have some components that are learned and some that are instinctual. At least in the case of the pheasant and other gallinaceous birds, the ability to peck is innate but what to peck at must be learned by watching what the parents eat. Oystercatchers obtain and eat the soft bodies of mussels in one of two ways. Some birds hammer a hole in the shell to get inside and some pry a gaping shell apart and cut the adductor muscle which holds the shells closed. Both methods are utilized in oystercatcher populations, but any one individual will only use one method. Is this learned or instinctive? To find out, birds using one method were given eggs from the nest of a pair that used the other method and vice-versa. If this feeding pattern is instinctive, then the chicks should behave as their true parents and not like their foster parents. In all cases, the chicks followed the pattern used by their foster parents, indicating that, at least in oystercatchers, foraging patterns are learned.

**Birds! From the Inside Out**

## Chapter 6 – Feeding Behaviors

All animals learn simple tasks but are birds capable of complex thought processes? We know many mammals, especially primates, are intelligent but how should we rate birds? These are not easy questions to answer and there is still much that is not known, but birds are often considered far less intelligent than their true abilities warrant. Intellectual abilities certainly vary between different families of birds and the answer to the question of how intelligent a bird is will depend upon what group you are talking about. Among the most intelligent of birds are the parrots and the corvids (jays, crows, ravens, nutcrackers and magpies). Different corvids have exhibited some remarkable behaviors both in laboratory situations and in the wild. American Crows will often use stones to help break open hard-to-crack nuts, but a particularly inventive one was once observed to place nuts upon the roadway, then stand to one side and wait until the nuts were cracked by passing cars. In another case a captive crow was fed a mash which was moistened with water to soften it before the bird ate it. Once, the human caretakers forgot to moisten the food before giving it to the crow. Undeterred, the crow picked up a cup, dipped it into some water and carried it over to the food where it poured the water on the food. This was not a behavior that it was trained to do. Crows have been observed to pull up fishing line that ice fisherman

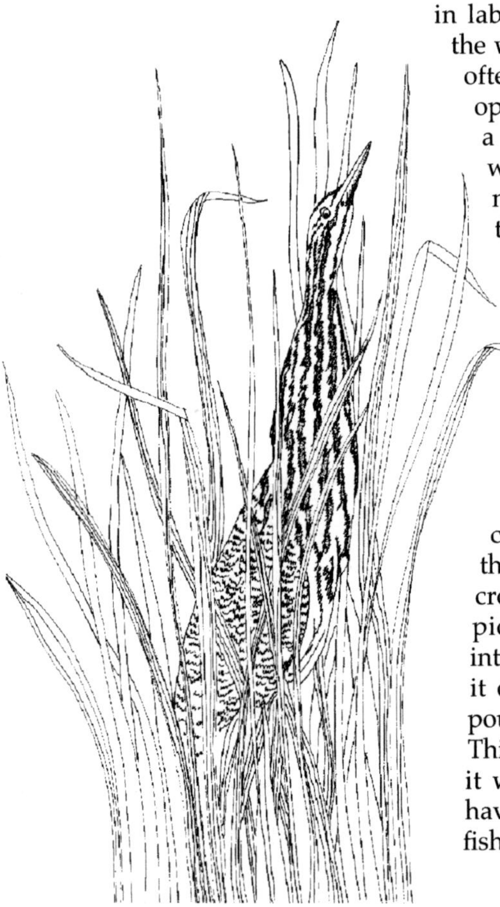

*Figure 6-4. A Bittern keeps its eye on potential danger by pointing its bill up, allowing the eyes to look forward. An unusual design, it also makes the bird "disappear" when one is looking for it in a marsh!*

have left dangling through holes in the ice. The crows apparently observed that there was food (fish or the bait) at the end of the line. A bird would grab the line in its beak, walk backwards and raise the fish or bait to the surface. But sometimes, the line would be long, requiring the bird to walk backwards a greater distance. This meant that by the time the bird let go of the line and rushed forward to grab its reward, the desired morsel had already sunk back into water beyond the bird's reach. After only a few such trials, the birds quickly learned to walk on the line, thereby preventing it from dropping back into the water.

Armed with these examples, perhaps the next time someone uses the phrase "bird brained" you'll have a ready retort, accompanied by a recitation of one of these examples, to help others appreciate the brains of birds.

*Figure 6-5. A Reddish Egret shades the water below to help it see its prey more clearly. An amusing (to us) dance can sometimes be observed as the bird tries to move closer to its potential meal.*

## Chapter 6 – Feeding Behaviors

# Chapter 7
# Bird Senses

Like all animals, birds obtain information about their environment through a variety of sensory stimuli. Of these, sight and sound are generally assumed to be the most important. There are no blind species of birds as there are of many other groups of animals. Song and other vocalizations are a key component in the behavior of many birds. These vocalizations serve many purposes and birds have a keen sense of hearing to receive them. We are beginning to understand, however, that other senses also play a key role in the daily life of a bird.

## Vision

It has long been assumed that birds have extraordinary visual abilities. We have all heard about hawks that can spot their prey at great distances, far exceeding human capabilities. Swifts and swallows darting about in the air easily spot tiny, flying insects upon which they prey. The marvelous assortment of patterns and colors found in the plumage of birds indicate that color must be an important component of their visual world.

Vision is an important aspect of a bird's total sensory perception, and all birds have well-developed eyes. The eyes of a living bird do not seem particularly large when observed in the field, but examination of a skull will reveal very large orbits (eye sockets) to accommodate equally large eyes. For many songbirds, the weight of the head is about one tenth the weight of the body, roughly equivalent to human proportions. In humans, the eyes weigh less than 1% of the total weight of the head. But in songbirds, the eyes may account for as much as 15% of the total mass of the head, and the eyes of some owls and eagles are as large as or larger than human eyes! The largest eyes of any living land vertebrate are found in the Ostrich, which has an eye 50mm (~2 in.) in diameter. Large eyes offer birds the advantage of larger and usually sharper images. The trade-off, however, is lack of room in the skull for the muscles that move the eye and, consequently, birds have a very limited ability to move their eyes. Most birds have some ability to direct the eyes forward, but the eyes of owls, for example, are fixed and immobile so that they must instead rotate their heads to look in different directions. Contrary to some popular ideas, an owl cannot rotate its head a full 360°. However, with the aid of a very flexible neck, the head can be turned about 270° before the owl has to turn its head back around

## Chapter 7 – Bird Senses

Figure 7-1. *A Common Snipe has eyes placed toward the top of its head, enabling it to see clearly overhead.*

to once again pick up a view of an object slowly circling it.

The vast majority of bird species have eyes placed on the side of the head, allowing extensive peripheral vision but very limited binocular, or stereo vision such as humans depend upon. When examining an object, most birds will use just one eye. The lack of binocular vision means that there is a loss of depth perception such as we use. To help overcome this, many birds (especially many chicks) bob their heads forward and backward, providing two quick views from different angles. In binocular vision, these two views are perceived simultaneously by each of two eyes, viewing the same scene from slightly different angles. Owls, with their forward looking eyes, have very good binocular vision, although the angle of view is slightly narrower than that of human stereo vision.

There are many interesting examples of birds whose eyes are positioned atypically. Bitterns have eyes that are rotated slightly downward. When alerted to potential danger, the bittern's concealment pose is to extend its neck and point its beak straight upward. This position makes the bittern, with its slender neck and streaked plumage, difficult to see against the reeds of its marsh environment. With the beak pointed skyward, the eyes, positioned low on the head, are now looking forward, providing the bittern with a stereo view of the potential threat. Wilson's Snipe, a bird of marshes and nearby tall grasses, has eyes placed somewhat more toward the top of its head, providing binocular vision nearly straight up. The American Woodcock has eyes that are located toward the top and rear, allowing for its best binocular vision to be rearward.

In our speech, we sometimes refer to eyes as "eyeballs" and thus have the impression that, like a ball, eyes are spherical in shape. Mammalian eyes are generally spherical or somewhat oblong in shape, but the shapes of birds' eyes are variable and differ between different groups of birds. The rear portion of a bird's eye is usually spherical, but the front half may be either somewhat flattened, well-rounded or nearly tubular in shape as is found in owls.

## Chapter 7 – Bird Senses

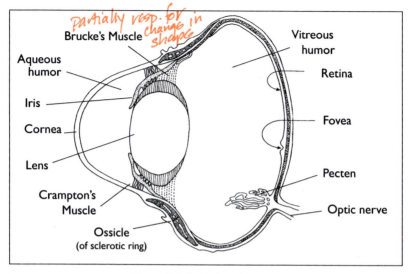

*Figure 7-2. Bird Eye Anatomy*

The basic anatomy of the avian eye is similar to that of a mammal but with some reptilian features still present. In particular, such structures include the **sclerotic ring**, overlying the outer portion of the eye, and the **pecten**, a highly vascularized structure projecting from the rear into the **vitreous humor**, the gelatinous main body of the eye.

The avian eye may be divided into two sections. The anterior portion of the eye contains the cornea and lens, and the larger posterior portion constitutes the main bulk of the eye. The sclerotic ring is the dividing point between the anterior and posterior portions of the eye. This ring is a series of 11 to 16 small, flattened bones, called **ossicles**, which form a circular plate around the lens and anterior portion of the eye.

Attached to these ossicles are two muscles, **Crampton's Muscle** and **Brucke's Muscle**. These help to hold the rather large eye in place within the skull, but their main function is to help change the eye's shape to allow focusing or accommodation. In the mammalian eye, only the lens shape is changed when focusing, but birds can change the shape of both the lens and cornea. The cornea's shape is changed when Crampton's muscle contracts.

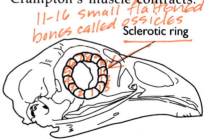

*Figure 7-3 Skull of chicken showing sclerotic ring.*

> **An Experiment to Try**
>
> The next time you have your eyes examined, try this simple experiment. The examiner will put drops in your eyes to dilate, or increase, the size of your pupils. The large pupil allows the examiner to look into your eye and examine the retina more easily. When your eyes are dilated, your vision becomes blurry because the depth of focus is very shallow. Now make a small hole in some cardstock with a pin. Notice that when you look through this small pinhole, your vision is much less blurry.

This changes the refractive ability of the cornea and lens and thus the eye is quickly refocused. Just behind the cornea is a clear, fluid-filled region called the **aqueous humor**.

Between the aqueous humor and the lens is the pigmented **iris** which gives the eye its color. Muscles associated with the iris control the size of the pupil. Pupil size (aperture) may be regulated independently in each eye, allowing a bird to have a large pupil in one eye while the other pupil remains small. For example, one eye might be focused on the ground, looking for food, while the other eye observes distance. Pupil size also controls the depth of focus, just as changing the aperture of a camera lens changes its depth of focus. The smaller the opening, the greater the depth of focus (also referred to as depth of field). In addition to regulating the amount of light entering the eye, pupil size, or a change in pupil size, may be used by some birds as a method of signaling others of the same species. This voluntary method of controlling pupil size is used by some parrots as a form of social communication.

The shape of the pupil is generally round, as it is in mammals. There are some interesting exceptions, however, such as the slit-like pupils of the skimmers (of the Family Laridae's sub-family Rynchopinae) or the nearly square pupil of some penguins (Family Spheniscidae), in particular, the King Penguin.

*Figure 7-4. Diving birds, such as cormorants, can quickly reshape the lens of the eye to focus at different distances.*

## Chapter 7 – Bird Senses

The lens of a bird's eye is softer and more pliable than the lens of a mammal's eye. Brucke's muscle is partially responsible for changing the shape of the lens in a bird's eye. The refractive index of a bird's cornea is nearly the same as the refractive index of water, so any change in the shape of the cornea has only a very small effect on the ability to focus underwater. As a result, most diving birds have a rather weak Crampton's muscle but a strong Brucke's muscle to quickly change the lens shape. This action is different than that of the human eye which has a firm, elastic lens rather than a soft, flexible lens. In the human eye ciliary muscles release tension on the lens allowing it to reshape itself. The bird lens is shaped by the muscles. Being soft, it is very quick to respond, allowing very rapid refocusing. This is a great benefit for a highly mobile animal seeking highly mobile prey, such as a swift in pursuit of a gnat or a kingfisher keeping a targeted fish clearly in focus throughout its dive. Most birds have powers of accommodation much greater than that of humans and are able to focus on even very close objects. Nocturnal birds are an exception and usually have accommodative powers much less than humans (even less than those of us who now need our bifocals to read anything held closer than at arms length).

When birds sleep, they close their eyes with their eyelids just as we do. Most birds raise the lower lid up and over the eye. Owls lower the upper lid, thus imparting a more "human" look from our anthropomorphic viewpoint. Birds also have a third eyelid called the **nictitans** or **nictitating membrane**. Owls have an opaque nictitans instead of the usual translucent membrane. In most birds, blinking is done with this membrane rather than with the eyelids. The epithelial cells lining the inner surface of the nictitans carry tears that are brushed over the surface of the cornea each time the eye is blinked. The nictitans has other functions as well. In many diving birds, the center of the nictitans is thickened and transparent, presumably acting as an additional lens in assisting these birds with visual accommodation underwater. Some birds use the whitish appearance of the nictitans as a method of signaling other birds. Members of the genus *Corvus* rapidly blink the nictitans during courtship or during aggression. The Black-billed Magpie, *Pica hudsonia*, shows the same behavior and actually has an orange spot on the nictitans that is a prominent visual cue.

Behind the lens, the rounded shape of the eye is formed by the gelatinous vitreous humor. As in reptiles, a bird's eye has a cone-shaped **pecten** projecting into the vitreous humor from the rear, near the point of attachment of the optic nerve. This pecten is simple in structure in reptiles but

*Cones: color perception & sharpness*
*rods: respond to low light levels*

## Chapter 7 – Bird Senses

often becomes quite large and more highly developed in birds, especially in diurnal birds. It is pigmented, highly vascularized with many tiny blood vessels (larger than capillaries which they are sometimes mistakenly called), and often very elaborately folded near its surfaces. The function of the pecten is still being debated among scientists, but the most widely accepted explanation is that it helps to supply the retina with nutrients and oxygen and aids in removing carbon dioxide since, unlike mammals, the retina of a bird has no blood vessels embedded within it.

The size of the eye, as well as certain behavior patterns associated with vision, suggest that birds have excellent vision. Examination of the retina further supports this assumption. The light receptive cells (cones and rods) are typically more numerous and more densely packed in birds' eyes than in the eyes of other vertebrates, including humans. Cones are the photoreceptors (light-sensitive cells) in the retina which are responsible for color perception and sharpness. The rods are another type of photoreceptor found in the retina. They are capable of responding to much lower light levels than are cones. The portions of the retina where the sharpest focus occurs, is a tiny, usually funnel-shaped depression known as the **fovea**. The fovea is usually densely packed with cones. Human eyes

*Figure 7-5. Swainson's Hawks have much greater visual acuity than do humans.*

have approximately 200,000 cones per square millimeter. Birds often have many more cones in the fovea of their eyes. For example, the Common Buzzard, *Buteo buteo*, of Europe, is reported to have over one million cones per square millimeter in the fovea of its eye! Even outside of the fovea, birds generally have more cones and greater visual acuity than do humans. In hawks, the upper surface of the retina often has a greater number of sensory cells. In flight, it is the upper portion of the retina which receives stimulus from the ground and potential prey, thus needs greater sharpness than does the lower portion of the retina which is receiving light from overhead.

Notice the behavior of a sitting Red-tailed Hawk, American Kes-

*diurnal birds have red, yellow, orange green oil*
*nocturnal are clear or pale yellow*

## Chapter 7 – Bird Senses

trel or other bird of prey when something else flies overhead. The hawk may turn its head upside down. This action allows the image to fall on the upper retina with its greater acuity, thus providing a clearer view.

For many birds, there is more than one fovea per eye. Diurnal birds that feed on the wing generally have a second fovea located more temporally (towards the outer side of the head) which may help with binocular vision. For many of these birds, there is a horizontal or **central streak** across the retina with a fovea at each end. This central streak is an area of greater visual acuity than the rest of the retina but less so than the fovea. This provides a quick way to scan across the horizon without the need to turn the head. The central streak is parallel to the horizon when the bird sits in its usual posture. Birds which have this central streak are all birds of open country which usually have a distant horizon in view. Forest-dwelling birds, where no long, distant horizon is typically visible, lack this central streak in the retina.

*low light*

Nocturnal birds have retinas with fewer cones but many more rods. Cones assist with color vision, which is much less important at night, but the additional rods allow greater sensitivity to lower levels of light. Many owls can see at light levels from one-tenth to one one-hundredth of that

### How well does an owl see?

Imagine four football fields placed end-to-end. You are standing on the goal line of the first field. A companion walks toward the far end. As she enters the fourth field, she continues walking until she reaches the 30 yard line. At that point she stops and lights a candle. With only the light of this candle, can you see a mouse running along that same 30 yard line? A Great Horned Owl probably can.

needed by humans. A single candle, placed at a distance of 1000 feet, provides sufficient illumination for some owls to see what is before them!

Well-developed color vision is achieved by the presence of numerous cones within the retina. Unlike mammals, birds have oil droplets associated with the cones, one droplet in each cone. These oil droplets are rich in lipids and carotenoid pigments. In nocturnal birds, these oil droplets are usually clear or pale yellow, but in diurnal birds the droplets are usually pigmented with red, yellow, orange and green. The function of these droplets is not yet fully understood, but it is thought that they may help filter some wavelengths of light and enhance contrasting colors or intensities of colors.

Birds can see beyond the normal

**Birds! From the Inside Out**

range of colors visible to humans. Although the receptors in the retina are sensitive to ultraviolet light, the lens of the human eye filters out these wavelengths, making it unavailable to us. The lens of a bird's eye does not filter out this light, and thus a bird can see the ultraviolet world invisible to us. Many flowers, especially night-blooming flowers, have patterns visible only in ultraviolet light. The patterns serve as nectar guides to night-flying insects like moths. These patterns are not useful to most birds, so how is ultraviolet perception of use to birds? We now know at least one interesting example, that of the Eurasian Kestrel, *Falco tinnunculus,* which hunts voles with the aid of ultraviolet light. Voles, like many mammals, follow routine pathways, marking them with urine, thus claiming territory and advertising sexual availability. But to the voles' great disadvantage, urine glows in ultraviolet light. These urine trails are visible to the Kestrel which, by following them, can more easily locate the vole, a common source of its prey-base.

The vision of hawks and eagles is often described as though their eyes were a set of miniature binoculars, magnifying the image many times and allowing the hawk to see its potential prey at great distance. The truth is not quite so simple. The fovea and central streak form a depression in the retina, allowing greater surface area for receptors in this region of the retina. In addition, the difference in the refractive index of the retina and vitreous humor bends the light rays before they strike the retinal surface. Together these features cause as much as a 30% increase in magnification of the image as it is projected on the fovea. However, this modest magnification is not entirely responsible for the keen eyesight of hawks and eagles. The increased density of receptors in the retina allows for a much greater **resolution**, or ability to discriminate tiny objects. Imagine two thin lines drawn very close to each other. As you move away from this image, the lines eventually appear to blend together and you are no longer able to distinguish or "resolve" them as discreet elements. At this point, if you only magnify the image, you will still not see two lines, just a larger image of the single, blurry line. The hawk, with its greater density of photoreceptors and increased resolution ability, would still be able to see the two lines long after we have lost the ability to discriminate such fine detail.

# Hearing

Hearing plays an important role in the lives of most birds. Birds sing to advertise for mates and announce territorial boundaries, calls provide communication between individuals and within a flock, and many birds use sound to help locate their prey.

## Chapter 7 – Bird Senses

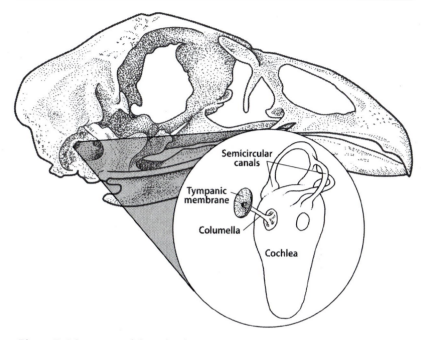

*Figure 7-6 Structure of the avian inner ear.*

Songs are learned and refined by listening to experienced singers of the same species. Potential predators may make sounds that alert a bird to their presence, and some sounds may even be used as navigational clues. Without sound, most individual birds would have a diminished chance of survival.

The anatomy of the avian ear has some similarities and many differences when compared to the mammalian ear. Bird ears, like mammals, are divided into three distinct parts: the external, middle and inner ear.

The pinnae, or outermost, visible portion of a mammal's ear is entirely lacking in birds. The outer ear of a bird consists of an open tube to direct the sound towards the middle and inner ear. The shape of this tube varies between different species of birds. It is shallow and weakly-developed in the Osprey, which does not depend on sound to locate its fish. Owls, such as Barn Owls, which depend upon sound to locate prey, have larger and more elaborate external ear tubes. Some birds have a dermal flap that can cover the ear opening or be raised to assist with hearing. Diving birds can cover the external opening to prevent water from entering. Owls also have a dermal flap that can be raised away from the external opening. The external "ear tufts" prominent on many woodland owls (such as Great Horned Owl or Long-eared Owl) are tufts of feathers that are not associated with the ears in any way. Instead of being atop the head, as these "horns" or "ears" would suggest, the real ears are located on the sides of the head, behind and slightly below the eye. Feathers smooth the contour of the head and completely cover the opening of the ears. The inner portion of the external ear terminates at the **tympanic membrane**, or eardrum, just as it does in mammals.

The eardrum defines the beginning of the middle ear. The middle ear of mammals is a cavity containing three small bones, the malleus, the incus and the stapes. Movement of these bones facilitates hearing. In birds there is only a single bone, the **columella**, which has the same derivation as the mammalian stapes. This tiny, column-like bone picks up vibrations from the tympanic membrane and transmits them to a membrane covering the pressure-sensitive, fluid-filled inner ear.

A bird's inner ear is much like our inner ear. Both have a tiny structure called the **cochlea**, which is lined with sensory cells. Each of these cells has small hairs that project into the fluid of the cochlea. As the eardrum vibrates, pressure waves are generated in the fluid-filled cochlea. These waves move the hairs, which in turn stimulate the sensory cells to respond, sending messages to the brain which interprets them as sound. The **semicircular canals** are also found within the inner ear. These small loops of bone serve as the organ of balance just as they do in mammals.

In many owls, the ears are not symmetrical in shape or location – one ear may be directed more downward. This asymmetry of shape and size may even be reflected in the shape of the skull in some owls, such as the Boreal Owl, *Aegolius funereus*. In this owl, the left side of the skull protrudes farther outward and downward than does the right side of the skull. This reflects the structure of the inner ear. This asymmetry in the ears of most owls provides them with a more complete sound picture, which helps them more easily locate prey by sound alone. Experiments have shown that a Barn Owl, *Tyto alba*, is able to capture moving prey in the complete absence of light, depending only on the sounds of the moving prey to serve as its guide. Diurnal owls and owls of open country typically have ears that are much more symmetrical, as they are more dependent upon sight for locating prey.

It has long been believed that birds have greater hearing abilities than do humans. More modern research is beginning to reveal that this is probably not true. Humans can generally hear fainter sounds than birds and birds appear to have a more narrow hearing range than do most mammals, including humans. A few birds (chickens, guineafowl and pigeons) are capable of hearing infrasound, frequencies below 20 hertz. These are extremely low frequency sounds and would include sounds generated by ocean waves, earth movements and other natural phenomena. Such low frequency sounds travel great distances and are possible to detect hundreds or even thousands of miles from their source. Imagine that a pigeon flying over the Oregon-Idaho border may actually be able to hear the Pacific Ocean 300 miles to the west! Such an ability could partially explain the well-known homing ability of some pigeons.

Birds do not hear ultrasonic sounds such as those used by bats when foraging for insects, but echolocation is known to be used by some birds. In Southeast Asia, some cave swiftlets navigate dark caves by emitting short bursts of sound. The Oilbird, *Steatornis caripensis*, of South America, a large, fruit-eating nightjar, also uses echolocation to navigate the shallow caves where it nests. Unlike bats, whose ultrasonic signals can discriminate small insects, the sounds of echo-locating birds are within the normal hearing range of both humans and birds. The

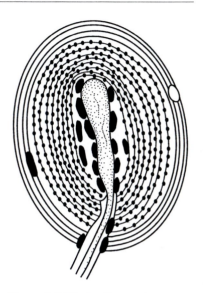

*Figure 7-7. Herbst Corpuscle.*

sounds returning to the birds allow them to navigate the wide corridors of the caves, but objects smaller than a few inches cannot be detected.

## Touch

A variety of stimuli provides a wealth of information necessary for the day-to-day existence of many birds. Mechanical stimulation (feather movement, touch, wind currents, etc.) provides information such as food location, barometric pressure, balance, wind speed and other cues necessary during flight. Much of this information is derived by stimulation of numerous **Herbst corpuscles**, the largest and most elaborate of a birds' tactile sensory receptors. Each corpuscle is flattened in shape and has a core

Figure 7-8. Dunlin, probing into the mud to taste for its food.

which is the enlarged end of a single sensory nerve cell. This swollen core is surrounded on either side by nuclei of axial cells which in turn, are surrounded by up to 12 layers (or lamellae) of connective tissue. Such corpuscles are very similar to the touch receptors (Pacinian corpuscles) found in mammals. They are found throughout the dermal layer of the skin and elsewhere, such as within the beak.

A bird's beak is not a hard, non-living structure similar to fingernails. Instead, it has many sensory cells within its structure which give the bird information about its surroundings. Birds such as snipe and godwits, which probe into the mud for food, have numerous Herbst corpuscles concentrated around the tips of their beaks. This allows the birds to discriminate between food and nonfood objects. Birds more dependent upon sight for food detection typically have fewer Herbst corpuscles and these corpuscles are more uniformly distributed along the length of the beak. Woodpeckers, which use their tongues to probe for insect larvae within the wood, have many Herbst corpuscles in the tips of their tongues. The follicles of filoplumes (see Chapter 1 on feathers) are surrounded by these corpuscles and give the bird information about wing position, air movements, etc., as the feather is moved.

## Taste and Smell

It has been demonstrated that many birds have an acute sense of taste, but it may be more narrowly focused than that of humans. Although similar in structure, birds have far fewer taste buds than do mammals. Humans have 9000 - 10,000 taste buds, rats have over 1200, but a Mallard has only 375 and a chicken has but 24. But even with fewer taste buds, some birds show a remarkable ability to discriminate food based on taste. Dunlins, *Calidris alpina*, and Sanderlings, *Calidris alba*, both search for their prey along open beaches or mudflats and locate prey visually and by taste. Experiments have shown that both of these sandpipers can distinguish between sand that contains prey species and sand from which it has been removed. Furthermore, unwashed sand with the prey removed is preferred to the same sand that has been washed as well as had prey removed. Apparently, there is enough residual taste left in the unwashed sand even though no prey remains.

## Chapter 7 – Bird Senses

Figure 7-9. Female Rufous Hummingbird feeding at a flower.

Hummingbirds are able to discriminate sugar concentrations as well as the types of sugar found in the nectars upon which they feed. A concentration of less than 1 part sugar to 8 parts water is ignored by hummingbirds. The Rufous Hummingbird, *Selasphorus rufus*, overlaps a portion of its range with Anna's Hummingbird, *Calypte anna*, but the Rufous Hummingbird likes sweeter nectars than does the Anna's. Consequently, these two species generally don't compete for the same sources of food, except where flowers are scarce or at hummingbird feeders where there may be vigorous competition for a single source. The sugar of choice for all hummingbirds is sucrose (table sugar), which is the sugar most often produced in plant nectars. Sugar mixtures containing some glucose are acceptable but less preferred than just sucrose, and mixtures containing fructose are even less acceptable. Feeders containing pure glucose (also called dextrose) or pure fructose are usually completely ignored. If you want to feed hummingbirds, refined, white sugar is the best (and most natural) choice of sugar to use. Unrefined cane juice, brown sugar and honeys are frequently ignored and are nutritionally not as useful to the birds. Honey can also contain fungal spores that can infect and potentially kill hummingbirds.

The ability to taste different substances varies greatly between species of birds. While hummingbirds show great sensitivity to sugars, many birds are completely indifferent to it, and instead are sensitive to other types of tastes: salt, acid or bitter. Jays and other birds violently spit out a Monarch butterfly. The larvae of Monarch butterflies feed on milkweeds, and thus incorporate the very bitter and poisonous alkaloids found in the milkweed plants. Other birds are insensitive to bitter tastes. This is especially true of ant-eating birds. (The sting of ants contains picric acid which is very bitter.) Capsaicins, the chemicals which make chili peppers so hot, may actually be attractive to some birds. Large amounts of capsaicins cause a burning sensation in the mouths of most mammals. Fruits, like chili peppers, that produce capsaicins are not eaten by mammals. Birds show no reaction to capsaicin and many birds will open chili peppers to consume the seeds. This is a benefit for the pepper plant. Seeds may be destroyed by passing through the digestive system of a mam-

## Chapter 7 – Bird Senses

Figure 7-10. Unlike most birds, Turkey Vultures have a keen sense of smell.

mal, but those same seeds are unharmed by a bird's digestive system. As the birds fly to other locations, the seeds are dispersed when the birds defecate.

While most birds have a sense of taste, it is often stated that most birds have little or no sense of smell. It would be wrong, however, to assume that the sense of smell is absent in the avian world. The majority of birds have small olfactory lobes in the brain, which lead many people to conclude that the majority of birds have little or no sense of smell. It was once believed that only birds with larger olfactory lobes (such as some vultures, kiwis, petrels and other "tubenoses") could possibly smell. Modern research is showing that assumption to be unfounded. We are now beginning to understand that most, if not all, birds use some smell in their day-to-day activities. During the mating season, some female ducks apparently produce weak scents from the preen gland. The male ducks are very sensitive to these scents and are attracted to females that produce them. Songbirds have very small olfactory lobes (even smaller than other birds) and were assumed to have no sense of smell, yet, it has now been demonstrated that European Starlings choose appropriate nest material based on smell.

The sense of smell is probably best known in kiwis and vultures. Kiwis, found only in New Zealand, are odd birds in many respects. The feathers of a kiwi are flexible and the barbs are hookless (see Chapter 1 on feathers) and have become more hairlike. Kiwis live in burrows, have small, weak eyes, are nocturnal and have a long beak used to probe into the mud for earthworms and other soil invertebrates. Unlike all other birds, whose nostrils are found at the base of the beak, the kiwi has nostrils at the tip of its beak which are used to search for food by smell.

Vultures have a reputation for being one of the few birds with a sense of smell, but Old World vultures lack a keen sense of smell and not all New World vultures depend on smell. The Turkey Vulture may stand alone in its ability to smell. Its large nostrils and large olfactory lobes enable it to be sensitive particularly to the smell associated with putrid flesh. A rotting carcass may attract Turkey Vultures from many miles away, but oddly enough, their sense of smell alone is not sufficient to locate prey. Once in the vicinity of putrid flesh, the vulture must find the source by sight, not smell. A carcass obscured from view will attract vultures, but they will continue circling, trying to locate the source visually. A common misconception is that Turkey Vultures "smell death" and circle a dying animal, awaiting its death. This is simply not true. A carcass only a few hours old that has not yet started to produce odors will not attract vultures.

A gas company in California once found Turkey Vultures to be very useful because of their sense of smell. Turkey Vultures are very sensitive and attracted to the odor of ethyl mercaptan. We can detect this smell as well and at very low levels. Since natural gas (methane) has no odor, ethyl mercaptan is added so that humans can easily detect the presence of methane in the event of a gas leak in our homes or offices. During the 1930s, there were some particularly troublesome leaks in gas pipelines of remote regions which were difficult to accurately locate. The gas company increased the concentration of mercaptan in the gas flowing through the pipelines of this region. As it leaked from the pipelines into the atmosphere, Turkey Vultures began to gather over the region of the leaks, allowing engineers the opportunity to locate the source of the problem.

# Chapter 7 – Bird Senses

# Chapter 8
# Avian Mating Systems

The need to reproduce is one of the most basic necessities of all life. Different forms of asexual reproduction are common in many single-celled organisms, in plants and in some multicellular animals. But sexual reproduction provides for the greatest genetic diversity which benefits a species as a whole. A great variety of mechanisms and behaviors have evolved to accomplish the exchange of gametes between two members of a species, usually one male and one female. (In some single-celled organisms, the two mating types are simply labeled [ + ] and [ - ] as both look and behave nearly identically.) As highly mobile creatures, birds have evolved some unique adaptations and behaviors for reproduction.

## Anatomy

Birds are highly specialized for flight. One important aspect of this specialization is the development of unique structures or methods of weight reduction. The reproductive anatomy of birds is therefore somewhat different than that of many other vertebrates. In most vertebrates, the gonads, or sex organs, are paired: two **testes** in the male and two **ovaries** and associated **oviducts** in the female (which lead into a single **uterus** and **vagina**). Both the left and

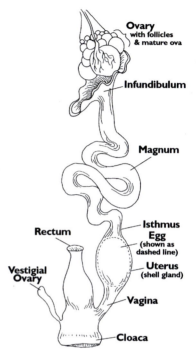

*Figure 8-1. Female Avian Reproductive System*

right gonad are functional and of equal or nearly equal size. In most birds, however, females have only one functional ovary and oviduct, usually the left one. The right ovary and oviduct typically degenerate early in embryonic development. Eggs, especially eggs near full development, add significant weight to the female and may impair her ability to fly. If both ovaries and oviducts were functional and contained mature eggs, the females of many species

## Chapter 8 – Avian Mating Systems

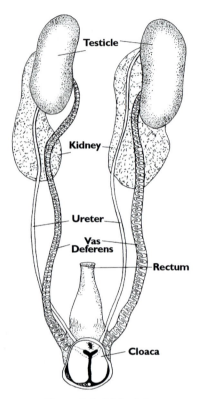

Figure 8-2. Male Avian Reproductive System

would be unable to fly until unburdened of one or more eggs.

The development of only one ovary and oviduct is not the pattern for all birds. Some raptors (in the genera *Accipiter, Circus,* and *Falco*) have two ovaries. Two ovaries occasionally occur in some individual females of other birds, notably gulls, pigeons and some songbirds. In most of these birds, even though two ovaries and oviducts may be present, it is still usually only the left ovary and oviduct that are fully functional.

The Brown Kiwi of New Zealand is one of the few birds that has two fully functional ovaries.

In males, both testes are usually developed and produce sperm, but the left testis is usually larger. The testes are contained within the body cavity and not held in an external sac, or scrotum, as in mammals. Maturation of sperm does not cause a significant increase in weight and so both testes are active. But the change in size of the gonads between breeding and non-breeding birds of either sex is very significant. A male bird in breeding condition may have testes that are up to 500 times larger than the testes of the same individual during the non-breeding season.

In a few primitive birds, such as ostriches, tinamous, curassows, and most ducks and geese, the males have an erectile, grooved, penis-like structure which is actually a modification of the cloaca wall. Turkeys and chickens also have a small penis, but in the vast majority of bird species males lack a penis entirely. The sperm is passed directly from the cloaca of the male into the cloaca of the female. This occurs when the male mounts the female and presses his swollen cloaca (known as a *cloacal protuberance*) against the vent of the female's cloaca during what is termed a "cloacal kiss." As there is no penetration by the male, it is usually necessary to have a series of these "kisses," each typically lasting one to two seconds, to accomplish adequate sperm transfer. This usually requires that both male and female be ready and willing for mating. In some birds, such as ducks, males may forcibly take females whether they are willing or not. The mating is then termed, appropriately, a rape. Some swifts apparently are able to mate while in flight.

Females may engage in several different matings, often with different males. The eggs that are fertilized will usually be the result of the last mating. It is thought that at least in some species, males may remove sperm stored in the female's cloaca from a previous mating. Male Hedge Accentors peck at the female's cloaca just prior to copulation. This stimulates the female to eject a small packet of stored sperm from her cloaca.

Eggs are usually fertilized within a few hours up to one or two days after mating, but the females of some birds can store sperm as long as 10 weeks after mating. Once sperm is released into the oviduct, it takes about 30 minutes to reach the **infundibulum** (the funnel-shaped end of the oviduct surrounding the ovary). Fertilization usually occurs in the portion of the oviduct just below the infundibulum.

## Mate Attraction and Mating Systems

In order to begin the breeding process, an individual must first obtain a mate. The majority of birds do not mate for life, so a new mate must be found in each breeding season. Birds use colors, songs, unique displays or a combination of these to seek a mate. Females have many different criteria they use to select the male of their choice. House and Purple Finches prefer males that are brightly colored and not dull, orange or yellow. Female Marsh Wrens judge the quality of their mate by the territory he defends.

*Figure 8-3. Male Sage grouse display for females on leks.*

## Chapter 8 – Avian Mating Systems

If less than 50% of a male's territory contains emergent marsh vegetation, no female will select him no matter how well he sings and displays. Many females judge the quality of their mate by the number of parasites that he has. A male with few parasites is a better mate choice. One notable example of this is the Greater Sage-Grouse. The most experienced males display in the center of the lek. Females generally ignore the less experienced males at the periphery of the lek and examine the inflated bladders on the breasts of the older males at the center. It seems that the females are not looking for the best display, but rather they are examining the bare skin of these bladders for the presence of reddish spots. Red marks indicate the presence of lice; females avoid such males in favor of males with clear skin.

Female Barn Swallows also assess the parasite load of potential mates. Females are most attracted to males with the longest tails. Shorter tail feathers during the breeding season are the result of damage done by feather lice. A longer tail usually indicates a more vigorous and parasite-free male.

There are a variety of mating systems generally used by birds worldwide. These various methods are summarized as follows.

*Figure 8-4. Sandhill Cranes, in a monogamous type of mating system, dance during part of their courtship rituals.*

## Chapter 8 – Avian Mating Systems

### ✻ MONOGAMY
**One male paired with one female during the breeding cycle**

Over 90% of all birds exhibit this type of mating system. For many birds, the two individuals may remain paired during most of the breeding season. Some species pair for a few years, some for life. Many species of birds, such as ducks and many songbirds, are not especially "faithful" to their mates during the breeding season and frequent extra-pair copulations may occur. In some birds, the pair-bond does not last much past egg-laying. Male ducks usually abandon their mates sometime during the incubation period. Geese, on the other hand, frequently form life-mates and males remain with their mates to help raise the young.

### ✻ POLYGAMY
**Two types: Polygyny and Polyandry (Classic Polyandry and Cooperative Polyandry)**

### POLYGYNY
**One male paired with two or more females during breeding cycle**

Only 2% of all birds are polygynous as a regular breeding pattern. In some species, the male defends the breeding territory but plays little role in the care of the young, as with many blackbirds. Males of other species, such as wrens, assist with parental care (incubation and/or feeding of

*Figure 8-5. Male Marsh Wrens sing to defend territory and attract mates.*

young) to some degree, often more with the primary female than with any secondary female. The most vigorous males usually hold the best territories. Less able males hold territories of lower quality and, as a result, may only be able to attract a single female. Some individuals of a predominately monogamous species may occasionally exhibit polygynous behavior. This most often occurs when food is abundant. Chipping Sparrows and Tree Swallows are examples of such species. If food suddenly becomes less abundant, the male frequently abandons the second female and her brood.

**Birds! From the Inside Out**

## Chapter 8 – Avian Mating Systems

Figure 8-6. *The Spotted Sandpiper has a system of serial polyandry in its breeding cycle.*

Many Icterids (New World blackbirds; i.e., Red-winged Blackbird, Yellow-headed Blackbird, meadowlarks, orioles, etc.) and Marsh Wrens are usually polygynous.

### CLASSIC (or Serial) POLYANDRY
**One female paired with two or more males during the breeding cycle**

Only a very small number of birds exhibit this type of breeding behavior (less than 1%). It occurs most commonly in some shorebird species, both species of buttonquails, both species of roatelos and in some rails. In classic polyandry, the females typically court the males. Males incubate the eggs and raise the young without direct help from the female, although the female often defends the males and the nesting territory from intruders.

After a clutch of eggs is laid, the first male begins incubation while the female seeks the attention of another male to care for another clutch of eggs. Sometimes as many as four separate clutches are laid. This is the typical pattern for the Spotted Sandpiper. Phalaropes are polyandrous, and simply show a pattern of sex-role reversal wherein the female is the more colorful bird and does the courtship while the male does all of the incubation and rearing of the young.

### COOPERATIVE POLYANDRY
**Two or more males cooperate to assist a single female**

This type of behavior is found in only six species of birds. They are: Galápagos Hawk, Harris' Hawk, Dusky Moorhen, Tasmanian Native-Hen (a type of gallinule), Acorn Woodpecker and Hedge Accentor (a songbird). In this type of arrangement, two or more males help a single female to raise a brood of chicks. In some cases, only one of these males is the actual mate while the others help with food gathering and care of the young. In the American Southwest, Harris' Hawks have a family arrangement of two males and one female at each nest.

# Chapter 8 – Avian Mating Systems

Figure 8-7. *Rufous Hummingbirds, like other hummingbirds, have a promiscuous mating system.*

## PROMISCUITY AND LEK BEHAVIOR
### No pair-bond formation

In birds, promiscuity is a mating system (not a moral judgement) whereby males display (using sound and ritualized activities) in an attempt to attract and mate with as many females as possible. No pair-bond is established between the sexes. The male's only role is to mate. In some species, such as hummingbirds, the males fly elaborate display patterns alone in an attempt to attract the attention of females and obtain frequent matings. In other birds, especially some grouse, males display on a lek or mating ground. On the lek, several males gather and begin a contest of elaborate displays. Females come to the lek and select a male that they deem most fit. In many cases, it is only the older and more experienced males that will be selected for mating.

There is a seemingly endless variety of behaviors associated with mating and mate selection in birds. Ninety to ninety-five percent of all birds are monogamous, yet each species will have its own unique pattern and signals. Students of bird behavior, whether beginner or professional ornithologist, will always be rewarded with something new if they are willing to take the time to carefully observe rather than simply identify a bird.

# Chapter 8 – Avian Mating Systems

# Chapter 9
# Brood Parasitism

Birds may be unique among vertebrates in terms of the amount of time and energy devoted to building nests and raising young. So it is only among birds where there exist so many opportunities to exploit this time and energy. Brood parasitism is probably the best known example of such exploitation.

An avian brood parasite is a bird who lays its eggs in the nest of another bird (most often, another species) for that host bird to incubate and raise the parasite's young. Brood parasitism is almost unknown throughout the rest of the animal kingdom, except in some insects and in a few species of fish. Amphibians and egg-laying reptiles typically leave their eggs unattended and usually provide little or no parental care. Thus, there is no advantage for these species to become brood parasites since neither parasite nor host provides care. Among mammals, whose young are all live-born, there exists no counterpart to this behavior.

There are over 9,600 species of birds worldwide but only a few more than 1% are known to be brood parasites. Conspecific (or intraspecific) brood parasitism (where eggs are laid in a host nest of the same species) has been reported for more than 50 species of birds. Of these, more than 30 commonly occur in North America. These conspecific brood parasites are known as **facultative brood parasites**, that is, brood parasitism is an option or done occasionally. None of these species depend upon brood parasitism as a normal method of reproducing.

Perhaps the best known brood parasites are the **obligate brood parasites**. For these birds, there is no other way of life. No obligate brood parasites ever build a nest or raise their own young. Worldwide, fewer than 100 species of birds are obligate brood parasites. These include all of the honeyguides (11 species), nearly half of the 130 species of cuckoos, two genera (*Vidua* and *Anomalospiza*) of finches, five (of the six) cowbirds and one duck. The most well known of these (at least in the popular press) are the Brown-headed Cowbird, *Molothrus ater* of North America and the Common Cuckoo, *Cuculus canorus* of Europe and Asia.

Brood parasitism offers some obvious advantages. The adult has to expend no time or energy to incubate eggs or to brood and care for hatchlings and later, fledglings. There are however, trade-offs. Considerable time

# Chapter 9 – Brood Parasitism

*Figure 9-1. The Brown-headed Cowbird is an obligate brood parasite.*

must be spent, especially by the female, in searching for nests of suitable hosts and waiting for just the right moment to approach the nest and lay an egg. This process must be repeated every couple of days. A female Brown-headed Cowbird is capable of producing about 40 eggs each breeding season. Once the eggs are laid, there are additional risks to the parasitic species. Many birds will reject any foreign egg found in their nest or simply abandon the nest and begin a new clutch. Yellow Warblers will often build a new nest right on top of the existing eggs. If a cowbird lays an egg in the new nest, the warbler may build on top once again. As many as four to five layers of nests have been made by Yellow Warblers in an attempt to avoid cowbirds. In the end, the warbler frequently still raises only a cowbird. Cedar Waxwings will severely damage cowbird eggs in the process of trying to remove them from the nest if the eggs are laid during the time period in which the waxwing is still laying its own clutch. After about three days of incubation, the waxwing may accept a new cowbird egg. Even if the egg is accepted by a waxwing, once hatched the cowbird chick may often starve due to improper diet. For the first three days after hatching, waxwings feed their young a diet of insects. After the third day, the adults begin to provide mostly fruit. A cowbird chick needs insects to remain healthy and will usually starve on a diet of fruit. However, if food is in short supply, the hungry cowbird chick may demand so much food during those first critical days that the waxwing chicks may not get sufficient food, and may starve as well. Goldfinches and other Carduline finches also make poor hosts for cowbirds. These birds never provide insects to their developing young, and so the cowbird chick never receives the food (insects) it needs to survive.

The Brown-headed Cowbird was once a species of the North American prairies, where it followed bison herds and fed upon insects stirred up by the feet of the bison, hence the name "cow" bird. This wandering existence makes it more difficult to find a suitable nesting location and, thus, a parasitic lifestyle is of great advantage. The Brown-headed Cowbird has now spread throughout most of North America, taking advantage of human habits of clearing forests, building roadways, etc. As it has spread, it

*[handwritten at top: 200 species taken advantage of & at least 140 successfully raised cowbird young.]*

## Chapter 9 – Brood Parasitism

has found more and more hosts. More than 200 species have now been reported as hosts for cowbirds and at least 140 of those have successfully raised cowbirds to fledging. Many of these hosts did not coevolve with the cowbird and have not developed natural behaviors to defend themselves against cowbird parasitism. The cowbird is often most successful with smaller species that are unable to destroy the eggs. Yellow Warblers and Song Sparrows are the two most commonly parasitized species. *[handwritten: smaller birds]*

Not all brood parasites are generalists, using many different hosts as does the Brown-headed Cowbird. Some species have only a few hosts that they will parasitize. In Africa, different species of parasitic whydahs have different species of estrildine finches as hosts. The pattern of spots in the mouth of the finch chicks is unique to each species. The parasite chicks have similar mouth patterns. A chick in the wrong host nest will not have a pattern matching its nestmates and will not be fed. In Argentina, there are Screaming Cowbirds, *Molothrus badius*, and Bay-winged Cowbirds, *Molothrus rufoaxillaris*. The Bay-winged Cowbird is not a parasite. It seldom builds its own nest, preferring to take an abandoned nest of another species. It is parasitized by the Screaming Cowbird and is its only host.

Figure 9-2. Yellow Warbler, a frequent Brown-headed Cowbird host.

### Common Cuckoo *[handwritten annotation of heading]*

One of the best known of all obligate brood parasites, the Common Cuckoo, *Cuculus canorus*, lives in Europe. Like the Brown-headed Cowbird, this cuckoo never builds a nest or raises its own young. However, it has some specializations for this way of life that go beyond those of the cowbirds. The eggs of the Brown-headed Cowbird all look the same regardless of which female laid the eggs or which geographic region she was from. Very often these eggs do not match the host eggs in appearance. This is not the case for the Common Cuckoo, whose eggs generally do match the host eggs. Different females may select host species different from those chosen by other females and cuckoos in northern Europe may parasitize different hosts from cuckoos in southern Europe. Most often, the eggs of the cuckoo look very similar to the eggs of the host. An individual female is not able to change the degree of spotting or streaking, or the egg coloration of her eggs to match widely different

**Birds! From the Inside Out**

## Chapter 9 – Brood Parasitism

host eggs, but within the species there is a wide variation of egg coloration to match the different hosts selected by different female cuckoos.

The cuckoo egg usually hatches about a day earlier than the host eggs. Once out of the shell, this blind, naked chick takes action to assure its own survival. There is a depression on the cuckoo chick's back between the wings. The newly hatched cuckoo chick moves around in the nest until a host egg is seated within this hollow. Then it pushes upward until the host egg is forced over the edge and out of the nest. It continues this action until all of the host's eggs are eliminated from the nest.

The honeyguides of Africa are also brood parasites and, like the cuckoo, eliminate their host competition. Barbets and woodpeckers (close relatives of the honeyguides) are the host species. Newly hatched honeyguide chicks have sharp hooks on the end of both mandibles. Using these hooks, the naked, blind hatchling of the Greater Honeyguide, *Indicator indicator*, grabs its host nestmates and ejects them from the nest. Other species of honeyguides use these hooks to attack the host chicks until they have killed them all. The host adults will remove the dead chicks and continue to take care of the one remaining honeyguide chick. These hooks have no other purpose, and within a few days to two weeks they will fall off of the beak.

The eggs of obligate brood parasites are usually a bit larger and typically hatch one to four days prior to the host eggs. The parasite chicks may take action against the other eggs or nestlings, as noted above. But even if no action is taken, this larger chick is the first mouth that the parents see and frequently gets most of the food brought by the parents, leaving the host young to slowly starve. The shell of the parasite egg is usually thick, making it difficult for a small species of host to break and remove the foreign egg. Orioles rarely accept cowbird eggs laid in their nests but they are a larger, and stronger bird than warblers or finches. Orioles usually pierce the cowbird egg and remove it from the nest. American Robins usually do the same. But Yellow Warblers or Song Sparrows are not usually

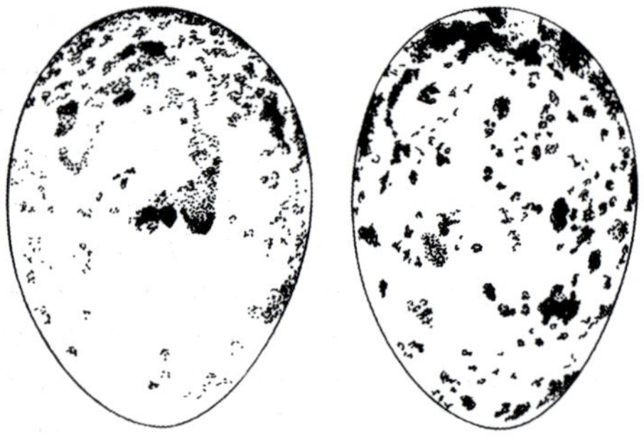

*Figure 9-3. The Common Cuckoo's egg (left) looks similar to its selected host (right), a Greater Reed Warbler.*

## Chapter 9 – Brood Parasitism

Figure 9-4. The American Robin usually is able to pierce and/or remove the egg of the Brown-headed Cowbird.

Figure 9-5. The Song Sparrow is a frequent Brown-headed Cowbird host.

strong enough to break or eject a foreign egg.

It is often assumed by birders that the presence of a brood parasite in the host nest will mean that none of the host nestlings will survive, but this is not always the case. For example, Indigo Buntings, when parasitized by Brown-headed Cowbirds, will usually raise some of their own chicks as well as the cowbird. However, there is some evidence that parasitized buntings may be put at a greater risk from predation of hawks or mammalian predators since the begging calls of parasite chicks are usually louder and longer than the call from an unparasitized nest.

In tropical America, nestling oropendolas sometimes have a greater chance of survival if a Giant Cowbird chick shares their nest. The Giant Cowbird, *Scaphidura oryzivora*, has only eight hosts. These include oropendolas, troupials and the Green Jay. Oropendolas construct large, pendulant nests and usually have two broods per season. As the female oropendola lays her first set of eggs, a Giant Cowbird may lay an egg in the nest as well. This egg is most often rejected by the oropendola. While the second set of eggs is being laid, the presence of a Giant Cowbird egg is often ignored and the cowbird is raised along with the oropendola young. Why reject the cowbird egg during the first brood but allow it to remain during the second? The answer is that Giant Cowbird chicks will eat botflies which are present during the period of time when the second brood is being laid and incubated, but not during the time of the first brood. Botflies are parasitic flies that lay their eggs on a living host. The botfly maggots feed on the flesh of the parasitized host, taking care not to destroy vital organs. The result is that the host is not immediately killed, and thus provides a living food source for the growing maggot. When the mag-

## Chapter 9 – Brood Parasitism

got is ready to emerge as an adult insect, the host is so weakened that it soon dies. Nestling birds parasitized by botflies usually die as they near fledging. Although oropendolas do not eat botflies, Giant Cowbird chicks do, and thus prevent the flies from laying eggs on any of the chicks. Oropendolas with cowbird nestmates therefore have a greater chance of survival. When the first brood hatches botflies are not present, so there is no advantage to the oropendola of the presence of cowbirds. Consequently, the eggs are usually rejected at that time. As an extra insurance against rejection, Giant Cowbirds in regions with no botflies lay eggs that are streaked and closely mimic the appearance of the oropendola egg. In the area with botflies, and a lesser chance of rejection, the cowbird eggs are usually plain white and easily identifiable from the host eggs.

*Black headed duck*

Another brood parasite that does not disadvantage its host is the Black-headed Duck, *Heteronetta atricapilla*, of central South America. This bird is one of the "Stiff-tailed Ducks" (Tribe Oxyur-ini) like the Ruddy Duck, *Oxymora jamaicensis*, of North America. It is the only known obligate brood parasite among all the world's waterfowl (and, interestingly, the only species of waterfowl where the female has a greater body mass than the male). The Black-headed Duck makes only one requirement of its host:

*(only obl. brood parasite among ducks (waterfowl))*

*Figure 9-6. In ducks, the Redhead is a facultative brood parasite, laying its egg in other duck nests about 40% of the time.*

incubate its egg alongside the host eggs until hatching. Once hatched, the young duckling quickly becomes independent of its host parents, leaving them free to raise their own brood. This is for the best for this duckling as the host species is frequently not another type of duck. Host species for the Black-headed Duck include: Black-crowned Night-Heron, Roseate Spoonbill, Southern Screamer, Coscoroba Swan, Limpkin, Spotted Rail, Maguari Stork, Chimango, Red-fronted and Red-gartered Coots, Rosy-billed Pochard, White-faced Ibis and Brown-headed Gull. Of these species, only the coots and Rosy-billed Pochard have been observed hatching young ducks. In the case of the coots, the ducklings hatch about two days prior to the coot eggs. While these coots are the most frequent hosts, a high percentage of the ducks' eggs never hatch, becoming buried too deeply in the coot nest for proper incubation.

Facultative brood parasites sometimes raise their own young, and sometimes resort to parasitism. Many ducks occasionally exhibit this behavior, but the Redhead, *Aythya americana*, a North American species, makes this a common practice. Redheads often build a nest and raise their own young just as other ducks do, but approximately 40% of the time Redheads employ brood parasitism, laying their eggs in the nest of other species, mostly other ducks. Canvasback, *Aythya valisineria*, is the preferred host. A Canvasback hen that serves as host to Redheads will raise some of her own young as well, but fewer than a hen in an unparasitized nest. Redheads will occasionally lay their eggs in the nests of several other species of ducks and are even known to lay eggs in the nests of coots or bitterns. But turnabout is fair play and Redheads have been recorded as playing host to eggs from several other species of ducks.

Ducks and many other species of birds are known to be occasional brood parasites, usually within their own species (intraspecific brood parasitism). Female Cliff Swallows, *Hirundo pyrrhonota*, occasionally remove one of their own eggs and place it in a neighboring nest within the nest colony. They carry the egg in their mouth, which is a remarkable task considering the egg's size, shape and smooth texture. The parasitized female accepts the egg as one of her own and usually lays one fewer of her eggs rather than have

Figure 9-7. *Cliff Swallow at nest. Females sometimes place their eggs in a neighboring swallow's nest.*

extra eggs. This type of parasitism most often occurs when nest sites are at a premium or when there is a shortage of food.

Incubation and the raising of young requires a large investment of time and energy on the part of the parent birds. Brood parasitism frees the adults from this time and energy expenditure, but the risks are high and only a small portion of a parasite's eggs will be accepted by a host. Therefore, the female brood parasite must usually produce a large number of eggs each season to insure that some will survive. It is human nature to make moral judgements, and so we all too often assess the parasites way of life as "bad." But we need to refrain from such judgements. What is detrimental to the host is certainly not detrimental to the parasite and, as we have seen, not all brood parasites have a negative impact on the host. Some are neutral and some actually aid the host as well as the parasite.

# Chapter 9 – Brood Parasitism

# Chapter 10
# Hatching and Care of Young

## Eggs & Egg Development

Avian eggs come in a marvelous assortment of sizes, shapes, and colors. Sizes range from the tiniest hummingbird eggs (some less than ½ inch in length) to eggs of the extinct elephant bird, *Aepyornis*. Fossil shells of this bird were fashioned into bowls by the early natives of Madagascar where these birds lived. The eggs were over 13 inches long and contained approximately two gallons of material. Usually, the larger the bird is, the smaller the egg is in relation to its body weight. Ostriches lay large eggs but each egg is less than 2% of the total body weight. The egg weight of a small passerine may be closer to 10% or more of the total body weight. The largest ratio of egg size to body weight are the kiwis of New Zealand. They generally lay only one egg per season, but that egg is ¼ of the body weight.

We are all familiar with the shape of a chicken egg. Most birds lay eggs that are more or less similar to that ovoid shape, but much variation occurs. Nighthawks lay eggs that are nearly equally rounded on both ends and look more elliptical. Many owls and some cavity nesting birds have nearly spherical eggs. The eggs are not likely to roll out of the deep nests of these birds. Cliff nesting birds, such as kittiwakes, lay eggs that are very pointed on one end, allowing them to roll in a very tight circle to avoid rolling off the ledge.

Eggs come in a variety of colors. Many species have white eggs with no pigment. Others have pigment uniformly distributed in the shell. Colors may be buff, brown, reddish, green or blue. Sometimes this color is laid down in streaks or splotches. Pigment glands in the wall of the uterus are responsible for secreting the pigments that result in spots or streaks on the outer surface of the shell. If the egg remains still in the uterus, the egg will have spots matching the location of the pigment glands. If the egg rotates while in the uterus, the eggs will appear streaked. The degree of streaking and swirling depends upon how much the egg moves over the pigment glands in the uterus. Pigments responsible for coloring egg shells come mostly from hemoglobin or from pigments of the bile that are the result of hemoglobin decomposition.

Nourishment for mammalian embryos comes across the placenta by way of the mother's blood. In birds, this nourishment

## Chapter 10 – Hatching and Nesting

is contained in the nutrient rich yolk. This yolk material is secreted by the ovary and walls of the uterus.

Egg formation begins in the ovary. Unlike mammals, birds have only one functional ovary, the right one is vestigial and undeveloped– an adaptation for weight reduction. When a ripe ovum ruptures from the ovary, it briefly enters the **infundibulum** (the funnel-like end of the oviduct surrounding the ovary) and remains here only a very short time (a few minutes). It then slowly passes along the oviduct, taking from less than one day to more than three days, depending upon the species. While in the oviduct, the albumen and most of the membranes are produced. One region of the albumen, the **chalaza**, forms a pair of twisted strands of albumen that attach to the yolk at opposite ends of the egg. The yolk is suspended by this chalaza in such a manner that the heavier vegetal pole of the egg is always down, regardless of the egg's orientation. This assures the embryo will be properly positioned as it develops within the egg. As the egg enters the uterus, the final constituents of the egg are completed and the shell (composed mostly of calcium carbonate) is secreted. The completed egg then enters the vagina and is soon expelled from the body. The journey from ovary to laying the complete egg is approximately 24 hours for most birds.

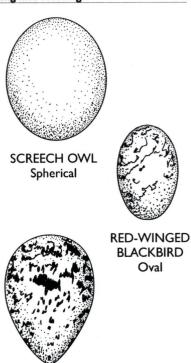

**SCREECH OWL**
Spherical

**RED-WINGED BLACKBIRD**
Oval

**KILLDEER**
Short Pyriform

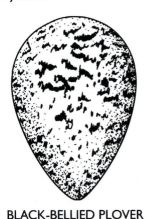

**BLACK-BELLIED PLOVER**
Pyriform

*Figure 10-1. A sampling of egg sizes, roughly proportionate to one another.*

## Chapter 10 – Hatching and Nesting

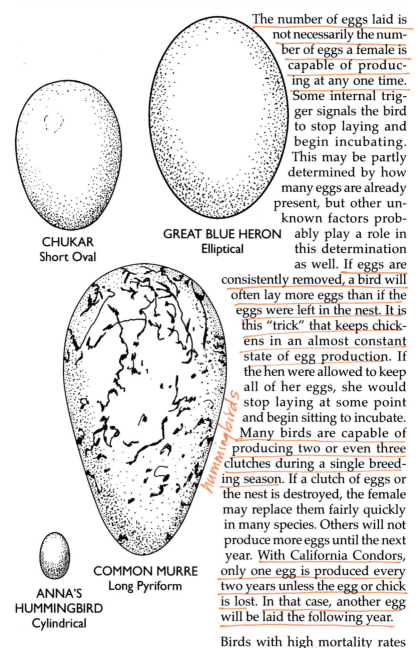

The number of eggs laid is not necessarily the number of eggs a female is capable of producing at any one time. Some internal trigger signals the bird to stop laying and begin incubating. This may be partly determined by how many eggs are already present, but other unknown factors probably play a role in this determination as well. If eggs are consistently removed, a bird will often lay more eggs than if the eggs were left in the nest. It is this "trick" that keeps chickens in an almost constant state of egg production. If the hen were allowed to keep all of her eggs, she would stop laying at some point and begin sitting to incubate. Many birds are capable of producing two or even three clutches during a single breeding season. If a clutch of eggs or the nest is destroyed, the female may replace them fairly quickly in many species. Others will not produce more eggs until the next year. With California Condors, only one egg is produced every two years unless the egg or chick is lost. In that case, another egg will be laid the following year.

Birds with high mortality rates usually produce a large number of eggs: California Quail usually

*Figure 10-2. A sampling of egg sizes, roughly proportionate to one another.*

## Chapter 10 – Hatching and Nesting

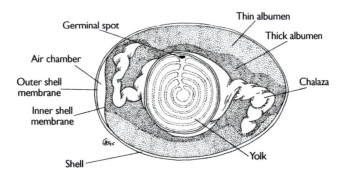

*Figure 10-3. Diagram of internal structure of typical bird egg*

lay 12-16 eggs. Many of these eggs and the newly hatched chicks will be lost to predators, disease or environmental factors. Of the chicks that survive to be adults, many will not survive to reproduce. A large clutch size assures that there will be many individuals so that a few will reach maturity and reproduce. The **precocial** (see page 97) nature of these birds means that the adults do not have to spend all of their time gathering food for a large number of chicks. For **altricial** (see page 98) birds, it may be a disadvantage to have a large number of eggs. Too much time would have to be spent in foraging for food for the chicks. The chicks would not be well fed and would be prone to weakness and disease.

## Incubation

Mammals have a gestation period where the developing young are nourished and maintained internally. The closest analogy for this in birds is incubation. Timings and patterns of incubation differ widely between different groups of birds. For instance, megapode parents do not incubate. Instead, the adults (usually the male) scrape together a large mound of decaying vegetation in which the eggs are laid. The males that build these mounds may spend more than six months building and tending them. As the vegetation decays, heat is produced. The male tends to this mound nest and periodically inserts his beak into the mound to sense the temperature of the egg chamber (to within 0.5°). He adjusts the temperature by either scraping more vegetation over the eggs to increase the temperature or scraping away some of the vegetation to lower the temperature. All other bird species must sit upon the eggs to incubate them.

Incubation may be the responsibility of only one parent or it may be a shared responsibility. A majority of bird species share incubation duties but often it is

## Chapter 10 – Hatching and Nesting

the female that has the greater responsibility. There are, however, examples of male-only incubation. In Spotted Sandpipers, the female courts the male and lays a clutch of four eggs. It is then up to the male to incubate and care for the young. The female seldom plays a part in this activity. She will lay four clutches of four eggs for each of four males. She will defend the territory of these four males but will usually not incubate or help with young. In the American Goldfinch, the female does all the incubation but depends upon the male for nearly all of her food during that time and later when the chicks hatch.

If something happens to the male, the female will have to abandon the nest. Doves share the responsibility and follow a surprisingly strict schedule. Females return to the nest at about 5:00 P.M. and remain on the nest throughout the night. Males relieve the females at about 9:00 A.M. the next morning and remain on the nest throughout the day. In many species, both sexes take turns incubating during the day but only the female incubates throughout the night. This pattern is reversed in woodpeckers, where it is the male that spends the night on the nest. Simultaneous incubation occurs in a small number of species. Some quail sometimes follow this pattern where the female makes two nests and lays two sets of eggs. The male incubates one set while the female incubates the other. Another type of simultaneous incubation is the pattern followed by anis. They cooperatively build a large nest in which many females will lay their eggs. Up to 60 eggs may be found in a nest. Several birds may be on the nest at one time to incubate these eggs.

For many birds, egg-laying may be separated by two or more days. Some birds begin incubation soon after the first egg is laid. In such cases, the

Hatching muscle

Egg tooth

*Figure 10-4. Hatching chick. Note egg tooth on upper mandible and hatching muscle on nape of neck.*

early embryo begins to develop and will ultimately hatch ahead of the later eggs. Barn Owls may lay as many as seven eggs over the course of 14-15 days. Incubation begins immediately, and thus the young birds in the nest will cover a wide range of ages. Competition for food by the chicks may be keen, and older chicks may attempt to dominate siblings. This pattern of **asynchronous incubation** and hatching can have severe disadvantages for many birds. For example, birds that are precocial, ground-nesting birds, such as Killdeer, would find it very difficult to feed and protect young birds that are out and running about while simultaneously continuing incubation of the unhatched eggs. All eggs must hatch at the same time to maximize chances of survival. To achieve this synchronous hatching, incubation does not begin until all of the eggs are laid. If incubation does not occur, the egg will remain alive but it will have an arrested development. Once incubation begins, all the eggs will develop at the same rate and hatch synchronously, often within minutes of each other.

## Hatching

The first struggle a bird must cope with independently is the long, demanding physical challenge of hatching. In the last days prior to hatching, the embryo has grown quite large and space is at a premium within the egg. When the hatching process begins, the embryo, which until now has had its head curled forward (tucking), pulls its head back and up, puncturing the membrane separating the contents of the egg from the air chamber at the blunt end of the egg. The chick then begins to lightly peck at the underside of the egg shell. This pecking is accomplished by contractions of the hatching muscle on the back of the neck. (After hatching, this muscle will weaken and become a vestigial structure.) This light pecking at the shell will eventually cause small fractures that will weaken the shell and eventually allow it to be broken. The hatching process may take two or three days before the bird is freed from the egg's confines. Most chicks turn within the shell while pecking with the beak. This causes a large hole to be fractured in the egg. Eventually the head is free and the bird then struggles to free its body from the egg. The feet are usually alongside the head while in the egg. At hatching, the feet extend forward helping to split the shell and free the chick.

The ability to fracture the shell is also aided by a specialized structure on the tip of the beak, the **egg tooth**. This hard, sharp structure is located on the tip of the upper mandible, just at the point where the tip curves downward. It is this egg tooth that usually helps to crack the shell. As the hatching muscle contracts, the head tips up and the egg tooth bangs against

# Chapter 10 – Hatching and Nesting

*Figure 10-5. PRECOCIAL: Killdeer chick*

the inside of the shell, causing small fractures in the shell. This action is referred to as "pipping." Once the bird has hatched, the egg tooth serves no other function and will be lost. Most songbirds gradually resorb the egg tooth. In other birds, it falls off, sometimes the same day as hatching, sometimes within two to three days. In some species, it may take two to three weeks to be shed. The megapodes, or mound builders, are the only birds without an egg tooth at hatching. The egg tooth develops early in these birds, but is resorbed before hatching, accomplished by kicking their way out of the shell with their feet.

## Hatching Patterns

**Superprecocial** — Hatchlings are completely independent of their parents. Megapodes or "mound builders" of Australia show this pattern of development. The hatchlings of these species are very well developed and leave the nest immediately upon hatching.

**Precocial** — Hatchlings are well developed and covered with down. They are alert with eyes open and ready to leave the nest shortly after hatching. Some follow their parents but are able to forage for themselves. This pattern is typical of most ducks and shorebirds. Some precocial hatchlings follow their parents and seem to know how to feed but not what to feed upon. They must be shown by their parents. Quail are one such example. Young quail will peck at almost everything, but must be taught how to discriminate between food (such as insects) and nonfood items (such as small stones). A few precocial birds, such as loons and grebes, follow their parents but do not feed themselves. For example, a

*Figure 10-6. SEMIPRECOCIAL: Gull chicks are semiprecocial.*

Common Murre chick, which also exhibits this pattern, will follow the adult male which brings fish to the chick. If something happens to the male, the chick will probably starve.

**Semiprecocial** — Hatchlings are down covered and have their eyes open. Many are able to walk a little but they remain in the nest and are fed by the parents. Gulls follow this pattern of development.

**Semialtricial** — Downy chicks are unable to leave the nest and are cared for entirely by their parents. Some, like herons, have their eyes open shortly after hatching. Others, like owls, hatch down covered but have their eyes closed.

**Altricial** — Hatchlings are naked and blind and completely dependent upon the parents for survival. They are able to gape and beg for food but are unable to leave the nest until more development occurs, taking from many days to weeks. Birds in the order Passeriformes follow this pattern.

*Figure 10-7. SEMIALTRICIAL: Burrowing Owl*

# Chapter 11
# Migration and Wintering

Each year, more than 200 species of North American birds migrate to tropical environments for the winter. Energetically expensive, this long-distance movement involves many risks. Despite much variability, in general the energetic costs of migration are greater for larger birds than for small ones. Conversely, larger birds are often able to survive colder weather more easily than small birds. Even though the benefits of migration outweigh the costs and risks involved, many birds do not survive its rigors. Some estimates suggest that as many as half of the migrants fail to return in the spring.

## Why Migrate?

The most obvious advantages of migration are securing a more favorable climate for the winter and a richer food source. Availability of food is probably a greater factor in determining the necessity of migration than is cold weather. If all birds that breed in northern latitudes remained throughout the winter, competition for food would be very keen and many species might find themselves without their primary food sources. While migration may reduce competition between some species, it is not a complete solution. Many birds, both migratory and non-migratory, find it necessary to change their diet for much of the non-breeding season.

## Dietary Changes

As the days grow shorter and colder, insect populations show very rapid declines. Species of birds that are obligate insectivores are forced to move to warmer climates where insects can still be found. This is the case for swifts, swallows, flycatchers and most warblers, all of which depend upon active, mostly adult flying insects. North American warblers are insectivores and most migrate south, but some also change their dietary habits during the winter months. Yellow-rumped Warblers, *Dendroica coronata*, are relatively common during the winter in many northern localities and as insects become harder to find, they begin eating some winter fruit. Many of these fruits have a waxy coating to help them last through cold

*Figure 11-1. The Yellow-rumped Warbler changes its diet during the non-breeding season.*

## Chapter 11 – Migration and Wintering

*Figure 11-2. Townsend's Warbler*

weather. Wax cannot be digested by most birds (or by most other animals), but Yellow-rumped warblers are one of the few species capable of digesting this wax and thus have a winter food source that is unavailable to most other species. In the western United States, Townsend's Warblers, *Dendroica townsendi*, are also found throughout the winter. They cannot digest wax and are forced to search for insect eggs, larvae and overwintering adult insects. This brings them down from the treetops to the ground and low bushes where they are often seen foraging during the winter.

A few other migratory warblers also change their diet during winter. Four species of North American warblers become nectar-eaters (at least partially) during the winter: Cape May Warbler, *Dendroica tigrina*, Orange-crowned Warbler, *Vermivora celata*, Nashville Warbler, *Vermivora ruficapilla*, and Tennessee Warbler, *Vermivora peregrina*. Competition for nectar is kept to a minimum by the fact that these warblers distribute themselves in different regions and habitats in winter. The Cape May Warbler is found in the Bahamas and the Greater Antilles. The Orange-crowned Warbler winters in the mountains of Mexico at high elevations. (Some individuals of this species remain much further north, especially during mild winters and, since nectar is not available, they continue to glean for insects.) Nashville Warblers are found at low elevation in the mountains of North and Central Mexico and in lowlands of the Caribbean. The Tennessee Warbler moves still farther south, dispersing from southwestern Mexico through Central America to Columbia.

Nectar serves as a food source for some other birds as well. In temperate North America, only hummingbirds and western orioles regularly eat nectar during the breeding season, but during the non-breeding season in their winter range, tanagers and eastern orioles may include nectar in their diets. Other birds that are resident or short-distance migrants in the tropics also depend upon nectar, but of all species of Neotropical migrants that regularly eat nectar, none cross the equator.

Other examples of species that may change their eating patterns during the non-breeding season include many of our northern-breeding flycatchers and vireos, which eat some fruit while they are in the tropics. Chickadees,

## Chapter 11 – Migration and Wintering

*Blue Grouse walk not fly upland in winter.*

primary insectivores during the breeding season, do not migrate and depend upon large quantities of seed to supplement their diet during the non-breeding season. They will readily come to a home bird feeder filled with black sunflower seeds during the winter and completely ignore the same feeders as spring and summer arrive. Chickadees and other insectivores (such as kinglets and creepers) are usually insect gleaners during the non-breeding season, searching surfaces of trees or bushes looking for insect eggs and larvae. A few species of insectivores (some flycatchers) are short-distance migrants, moving only to southern North America. Along streams in some of these areas, there are still some adult insects emerging from the water throughout the year, providing food for a few bird species.

The Sooty Grouse, *Dendragapus fuliginosus*, uses still a different strategy. It is not exclusively an insect-eater but does eat large quantities of insects, especially grasshoppers, during the summer along with some fruit, leaves and flowers. Young birds are fed almost exclusively insects. As winter approaches, Blue Grouse have an unusual pattern of movement, going from their low-elevation breeding grounds to high-elevation coniferous forests. As snow covers the ground, grouse are found in the conifers, where they eat the soft buds at the tips of the branches. It is interesting to note that Blue Grouse undertake this reverse migration by walking, not flying.

Strictly seed-eating birds have less need to migrate since seeds are often available throughout the winter. For instance, in the deserts of southwestern Mexico, many seeds are abundant during the winter. As a result, none of the strictly seed-eating birds migrate from this area. In many areas, local populations of birds may wander in search of the best seed crop. Crossbills are one such wandering species. They are found throughout the year in the coniferous forests in northern latitudes of the northern hemisphere, where they move through the forests in search of an abundant cone crop. When a rich source of seed-laden cones is found, the crossbills remain and begin their breeding cycle. While breeding

*Figure 11-3. Blue Grouse*

is more frequent in spring and summer, crossbills may breed at any time of the year. A good food source helps assure successful breeding even if the weather is cold.

*Figure 11-4. Red Crossbill, showing crossed beak*

## The Role of Weather and Temperature Regulation

Cold weather is responsible for many diminishing food supplies. Many insects may be found along streams and lake shores from spring through fall. As these waters begin to freeze over, the adult insects are no longer present and access to larvae under the water is cut off. Obligate insectivores, such as flycatchers, are forced to look elsewhere and move to warmer environments. If food is not a problem, most birds can endure short periods of cold temperatures, but not all birds can exist in prolonged cold. Some birds must gradually change their metabolic pathways to acclimatize as the weather becomes cooler. For example, American Goldfinches acclimatized to winter conditions can keep a constant body temperature and survive temperatures as low as – 90° F.

for 5-6 hours! These same birds during the summer would only be able to survive for about one hour in such extreme cold. As temperatures drop lower, birds eventually begin to shiver to maintain a constant body temperature. Most of this shivering occurs in the large pectoralis muscles of the breast. (This large muscle mass pulls the wings down and forms the largest muscle mass in flying birds — up to 15% of the total body mass.) In some birds, especially running birds, the leg muscles also shiver. Shivering produces heat but increases consumption of oxygen, requiring use of greater fuel reserves. Different species of birds begin shivering at different temperatures. Those species that spend their whole lives in northern latitudes begin shivering at much lower temperatures than birds of warmer climates. One of

*Figure 11-. American Goldfinch*

several factors that helps explain this is size. Large birds are able to maintain their body temperature much better than small birds. Smaller-bodied animals require a higher metabolic rate in order to meet the energy demands of their bodies. As size decreases, the ratio of body surface to body size increases and it becomes increasingly difficult to maintain a constant body temperature. There is a "smallest possible size" limit, and any animal smaller than this would lose more heat than it could produce through metabolic reactions.

In birds, this minimum size is about two grams (slightly less than the weight of a dime). The Bee Hummingbird, *Mellisuga helenae*, of Cuba is virtually at this limit. It cannot increase its metabolism sufficiently to allow it to survive in colder weather. During the cool evenings, many species of hummingbirds go into torpor to help survive the night. The Little Hermit, *Phaethornis longuemareus*, of Central America must feed every 3-5 minutes on the average to maintain its body temperature of ~100° F. To survive the night, it lowers its temperature to ~55° and its heart rate drops to just under 40 beats per minute. (Over 1000 beats per minute is normal for this bird in flight.)

Some hummingbird species do live in cooler environments. They generally are slightly larger hummingbirds than many in warmer

*Figure 11-6. Rufous Hummingbird*

areas. The Rufous Hummingbird, *Selasphorus rufus*, is found farther north than any other hummingbird, having a range which extends into southern Alaska. A particularly large hummingbird is found near the snow line at over 15,000 ft. elevation in the Andes. Anna's Hummingbird, *Calypte anna*, once a California resident, is now found throughout the year as far north as southern British Columbia. In fact, in this northern extension of its range, it is more commonly seen in the winter than during the breeding season. (This might be simply because it is more easily seen during the winter when it is readily attracted to – and sometimes needs – hummingbird feeders.)

## Cold Weather Survival Strategies

A few birds have developed very different strategies to survive cold weather. Ptarmigans and Ruffed Grouse, *Bonasa umbellus*, bury themselves in the snow by flying directly into a snowbank. This action leaves no tracks for a predator to follow and the snow provides an excellent layer of insulation to help maintain heat. Many birds seek shelter in cavities of trees or rocks. Old woodpecker holes are often used by songbirds for protection during a cold night. Rosy Finches have been found huddled together in Cliff Swallow nests. One observation was made of over 100 Brown Nuthatches huddled together in a single pine tree cavity [cavities of trees]. They were so tightly packed that some of the inner individuals had suffocated. Doves sometimes gather in a tight group of 10-12 individuals that forms a pyramid two to three rows deep.

Some mammals hibernate to escape the cold of winter. Hibernation in birds is completely unknown with one notable exception. The Common Poorwill, *Phalaenoptilus nuttallii*, has occasionally been found "hibernating" in the mountains of the Southwest. This "hibernation" is really a form of deep torpor rather than the true hibernation of small mammals. During this torpid period, a Poorwill maintains a body temperature of 41-43° F for a period of two to three months and decreases its oxygen consumption by 90%. A hummingbird or other animal in torpor will quickly become active as its body temperature is increased. On the other hand, an animal in true hibernation will not recover full activity until the end of a timed cycle (weeks to months). Poorwills found in deeper torpor will only recover to full activity after 6-7 hours of being at warmer temperatures.

Figure 11-8. Common Poorwill

## Migration Patterns

The patterns of migration differ from species to species. Buff-breasted Sandpipers, *Tryngites subruficollis*, breed in the Arctic and spend the winter in the pampas

Figure 11-7. Sanderlings on the beach

of Argentina. Sanderling, *Calidris alba*, which also breed in the Arctic, winter over a broad range of shorelines from the coast of northern Washington all the way to northern Chile. Some species are *partial migrants*, that is, not all subspecies or populations migrate. The White-crowned Sparrow, *Zonotrichia leucophrys*, is a partial migrant. The race which breeds along the California coast is resident there all year long. The race that breeds in western Oregon, Washington and British Columbia (primarily coastal) is a short-distance migrant. Races of White-crowned Sparrows found at higher elevations and those in northwestern Canada and Alaska are long-distance migrants.

Some birds show different patterns of migration within a species that are not necessarily racial, but may have more to do with geography. The House Sparrow, *Passer domesticus*, is not migratory in its native home of Britain and western Europe, nor is it migratory in the United States where it has been widely introduced. However, in China, where it has also been introduced, the House Sparrow is migratory. Some evidence suggests that the Barn Owl, *Tyto alba*, is migratory only in the most northern portions of its range. For some birds, migration is really movement over a short distance or expansion into habitats not used for breeding. Some birds that breed at higher elevations simply move

Figure 11-9. *White-crowned Sparrow*

to a lower elevation to spend the winter. Winter Wrens, *Troglodytes troglodytes*, typically breed in the dense understory of coniferous forests. During the winter, they can be common in hedgerows and brush thickets found even in some suburban areas. Those at more northern latitudes may migrate south into some southern portions of the United States where they do not breed.

Two western thrushes, the Swainson's Thrush, *Catharus ustulatus*, and the Hermit Thrush, *Catharus guttatus*, exchange seasonal habitats in the extreme western portions of their ranges. In the west, the Swainson's Thrush commonly breeds in the underbrush of coniferous and mixed woodlands and in brushy riparian habitats where its activities are generally confined to within ~ 20 feet of the ground. The Hermit Thrush may be found in some of the same coniferous forests, especially at higher elevations. Although the Hermit Thrush nests on the ground, most of its activities (foraging, singing, etc.) are carried out in the upper half of the forest canopy. During the winter months, the Swainson's Thrush migrates to forests from central Mexico all the way to Argentina. In the extreme western United States, the Hermit Thrush may spend much of the winter in habitats that were occupied by Swainson's Thrush during the summer. In the northern and interior portions of its range, the Hermit Thrush is migratory, wintering in Guatemala and El Salvador.

In some birds, males and females of the same species show different migration patterns. Dark-eyed Junco, *Junco hyemalis*, males generally winter farther north than do females. In many species of ducks, females return in the spring to the same region where they were hatched. Males simply follow the females and are not faithful to the site where they themselves were hatched.

Many birds migrate in a pattern of "leapfrogging." In this pattern of migration, individuals breeding the farthest north will migrate the farthest south, passing over those in the middle. In some cases, those in the middle of the range may not migrate at all or move only short distances. The Fox Sparrow, *Passerella iliaca*, along the extreme west coast of North America follow this pattern. Birds that breed in southern Alaska spend the winter in southern California.

Figure 11-10. *Swainson's Thrush*

The race breeding along the coast of northern British Columbia winter in western Oregon. The race which breeds around Puget Sound and southern British Columbia remain there throughout the winter.

For those species that do migrate to the tropics, the winter survival rate is generally higher than it is for species that remain in temperate regions. On the other hand, those that do not migrate from temperate regions often have higher reproductive success. For tropical residents, there is generally low productivity, but high survival. Fewer nests are successful and the clutch size is smaller than that of temperate habitat birds, but the adults are generally more long-lived. The average annual survival rate for resident tropical birds is 80-90%. The survival rate is 50% for migrants and 20-58% for temperate residents.

The process of migration is stressful and presents many obstacles

## Chapter 11 – Migration and Wintering

for birds to overcome. For any given species, the patterns and migration routes for spring and fall migration are often quite different. In general, it is the fall migration that provides greater barriers to surmount. In the fall, the food supply is rapidly becoming scarce for many species. It is essential to reach the wintering habitat as quickly as possible. If all migrating birds moved slowly overland together, the number of individuals would cause a demand for food that would far exceed the available supply. Many fall migrants make the journey in one long, nonstop trip, thereby eliminating foraging competition en route. For many eastern songbirds and shorebirds, this involves extended flights over the Atlantic Ocean rather than a more direct overland route. It is estimated that each fall, up to 12 million birds pass over Cape Cod and head southeast across the open ocean in the direction of Bermuda. When they reach latitudes of southern Florida (but still over open water) they encounter trade winds that turn them in a southwest direction toward the north coast of South America. Since most of these birds cannot land on water, this is a forced nonstop flight of up to ninety hours.

The fuel to drive these long migrations is fat. When metabolized, fat yields twice as much energy and water per gram than do carbohydrates or proteins. In birds, rather than being stored as inert, fatty deposits, the fat is stored in adipose tissue under the skin, in many muscles, in the peritoneal cavity and in some internal organs. Unlike humans and some other mammals, no fat is stored on the heart.

Prior to the fall migration, birds store large quantities of fat. Non-migratory birds have 3-5% of their body mass as fat. Birds which are short-range migrants (those which make short distance flights and stop to refuel) have 13-25% of body mass as fat. For the long distance migrants, the percentage of body fat is often highest just before migration. Dunlin, *Calidris alpina*, may have 30-47% of their body mass as fat prior to migration. Many small birds, like warblers and hummingbirds, nearly double their body weight before beginning a long, nonstop flight.

Many factors combine to determine what distance a bird is capable of flying nonstop: quantity of stored fat, ability to utilize this fat, speed of flight, and wing

*Figure 11-11. Fox Sparrow*

## Chapter 11 – Migration and Wintering

shape all play a role. A small bird, such as a warbler, with 40% stored fat should be able to fly about 100 hours. At the rate of speed that these birds can fly, that results in a distance of over 1500 miles. Shorebirds, which typically fly at faster speeds, can usually fly a distance of 1800-2400 miles. The Ruby-throated Hummingbird, *Archilochus colubris*, of eastern North America, migrates overland until it reaches southern Florida. From there it travels over open ocean across the Gulf of Mexico to the Yucatan Peninsula, a distance of over 500 miles. This seems impossible until you look at the energetics of fat metabolism. This species of hummingbird will store up to 2 grams of fat prior to its long flight. Flying at a speed of approximately 25 miles per hour, fat will be burned by the hummingbird at a rate that will allow a flight of over 600 miles — more than enough to complete this flight, especially since trade winds at that time of year are usually in the same direction that the bird is flying (east-to-west).

The spring migration pattern of most migratory birds is often quite different from the fall pattern and may take a longer time using an overland route. Most waterfowl typically migrate in stages rather than one long flight. While many birds make long, nonstop flights, others stop in key staging areas to rest and refuel for two to three days. This is particularly true for spring migration. The arrival of migrants on many of these staging areas coincides with a sudden increase of food supply. This is especially true for many shorebirds. Huge numbers of shorebirds arrive on the north Atlantic coast of North America at the precise time that horseshoe crabs come up onto the beaches to spawn. The horseshoe crabs lay their eggs in the sand and in a matter of hours, millions of these eggs become a valuable food source for vast numbers of shorebirds.

*Figure 11-12. Dunlin, feeding on small invertebrates*

# Chapter 12
# Attracting & Feeding Wild Birds
## Facts and Misconceptions

Outside the window a Spotted Towhee scratches for seed in the grass. Nearby, are three Dark-eyed Juncos and two Golden-crowned Sparrows. At the feeder overhead is a brightly colored male House Finch. These may be some of your rewards for attracting birds into your yard. But are you doing everything possible to assure the safety and welfare of the birds you attract?

The popular press is often filled with ideas about what to do and what not to do for birds in your yard. Some of this information is very good and some, while well-intended, is based on faulty assumptions and is not true. Let's first examine some popular misconceptions.

It has often been suggested that feeding birds may be detrimental to the local population by causing a dependence upon an unnatural food source or by causing delays in migration. There is no biological data to support either of these assumptions. A shortage of natural foods may cause birds to leave an area or move farther than normal. For example, a significant decrease in the lemming population on the Arctic tundra will usually force

*Figure 12-1. Black-capped Chickadees switch from a diet of insects in the summer to mostly seeds in winter.*

Snowy Owls to move south much greater distances than usual in search of prey. (Years with high owl populations usually coincide with low lemming numbers, until the lemming population experiences a crash, at which time owls must fly farther to feed.) Crossbill movements also depend upon food availability. Crossbills stay in flocks that wander throughout the coniferous forests. When they come upon a large cone crop with an abundance of seed, they settle into the area and often begin a breeding cycle. Although this breeding peaks during spring and summer, it may occur at any time of the year. A scarcity of wild foods may be a factor for movements in many species, but

## Chapter 12 – Feeding Wild Birds: Facts and Misconceptions

rarely could this be significantly offset by backyard feeding. Conversely, an overabundance of food will not delay migration. Departure and arrival dates remain relatively stable for most species, regardless of food available at feeding stations. Day length and hormonal changes within the birds play a much greater role in determining migration times.

Another common misconception is that hummingbirds will not migrate in time if you leave your feeder up too long. In fact, there is no data to support this claim. Hummingbirds which regularly come to feeders leave within the same time frame as those that do not regularly visit feeders. In the Pacific Northwest, you may have an additional treat if you leave your feeder up throughout the winter—an Anna's Hummingbird. Anna's Hummingbirds are now well established as winter visitors as far north as southern British Columbia, and some remain throughout the year as part of the permanent avifauna. (This should be expected as they are non-migratory throughout their range in California.) Winter food for Hummingbirds is sparse here, so if a hummingbird is visiting your feeder, be sure to keep it clean and filled with fresh sugar-water. In really cold weather, it may be necessary to have a heating device near the feeder. Some people keep two or more feeders in the house ready to use. As the outside feeder begins to freeze, a fresh one can be brought out to replace it. (Note: You can also keep a shop light nearby and place it where it can shine on the feeder in very cold weather, as we do. This keeps the water from freezing during extreme cold, at least on that side of the feeder.) Never put anything in the solution to prevent freezing. Such additives would almost certainly be harmful if not fatal to the birds.

There are some people who think that it is a bad idea to have hummingbird feeders at all. They reason that a hummingbird will be come "addicted" to the sugar-water and not feed upon enough natural foods. This is simply not true. A hummingbird's rapid metabolism requires a high-energy source of food. In the natural world, this energy source is largely sucrose (table-sugar) found in the nectar of many flowers. A solution of table-sugar and water (1 part sugar: 4 parts water) closely approximates the natural concentration of sugar in most nectars. Hummingbirds obtain the necessary dietary protein by eating small insects and spiders. A hummingbird will readily abort sipping from a feeder if a small insect suddenly ventures nearby.

Some people, unaware of the dangers, advocate the use of honey in hummingbird feeders. Don't ever use it! Many people mistakenly believe that whatever is good (or bad) for humans is good (or bad) for all other animals. We

## Chapter 12 – Feeding Wild Birds: Facts and Misconceptions

all know that too much sugar is bad for us and many believe that honey is better. But honey is made of sugars as well. It is just that most of the sugars in honey are not sucrose (which is usually the only sugar we refer to when we say "sugar"). Sucrose is the naturally occurring sugar in nectars and thus is the sugar that a hummingbird needs. Honey (which is not nectar but a product manufactured by bees) has been shown to be often quite harmful to hummingbirds. It is not as readily digestible to hummingbirds as sucrose and frequently contains spores of a fungus that can infect the tongues of hummingbirds, making it impossible for them to feed. Any artificial sweeteners are also to be absolutely avoided. They provide the hummingbird with no calories, and thus the hummingbird has to expend even more energy in search of proper food.

Once you have the proper sugar-water solution in your hummingbird feeder, leave it colorless. Hummingbirds are more attracted to red than to any other color but a red feeder or red on the base of the feeder is more than sufficient to attract hummingbirds if they are in your neighborhood. Red food-coloring has no positive value for the birds and, while no harm has yet been established, it is best to take no chances. Hummingbirds are very small and need to feed quite frequently. If much of their diet is artificially colored sugar-water, then they often can ingest very large quantities of red dye (in proportion to their body weight). Sometimes, they have taken in so much red dye that their fecal droppings are red. Such concentrations of red dye cannot be too healthy to the bird.

Are any other commonly provided foods harmful to the birds we are wanting to feed? Generally, no. However, there are a few things to be careful about. Fresh peanut butter is quickly taken by over 50 species of birds throughout North America, but a few studies have shown that some birds occasionally have difficulty swallowing it. The peanut butter becomes lodged in the throat and the bird suffocates or is unable to swallow. Not all studies have found this result, but it's best to play it safe. Don't use just raw peanut butter. Mix it with cornmeal to make a mixture that is not quite so sticky. Bread crumbs, while not the most nutritious of foods for many birds, are readily taken by several species. Use dry bread, and avoid using fresh, doughy breads that could form a sticky ball (much like the peanut butter). Is there any problem with bird seeds? Most commercial mixes are quite good and offer no direct problems. You should, however, take care that the bird seed you provide will remain relatively dry until it is eaten. Moist grain in a feeder will often become moldy and some of these molds are harmful to the birds.

## Chapter 12 – Feeding Wild Birds: Facts and Misconceptions

*Aspergillus fumigatus* is a common mold on wet grain which can infect the respiratory system of birds. If the seed in your feeder is damp and swollen, remove it, clean your feeder, and replace the seed with new dry seed.

Lately, it has become popular to use bird seed instead of throwing rice at weddings. The suggestion is that rice will be eaten by birds and then swell in the stomach, causing problems for the bird. An interesting idea but completely wrong. The birds which might eat the rice grains would be seed-eating birds. Seeds are not always easy to open and grind for digestion. Such birds usually have rather formidable gizzards for very effectively grinding of these seeds. Rice grains would easily be ground in the process. Even if this were not the case, there would still be no harm. Cattle and horses continually rechew their food, and thus have it in their system for hours. Birds are just the opposite extreme. Birds have a very high metabolism and the food that is eaten is very quickly digested and excreted (in some cases, as quickly as twenty minutes). There simply isn't enough time for rice grains to sit in a bird's stomach and swell. And even if digestion was slow, there is yet another fact to disprove this fallacy and this one you can easily demonstrate for yourself. Put some rice grains in water and watch what happens. Do they swell? No—not even

hours later. Now heat the water to about 105° F. (This is close to the body temperature of most small birds). Now what happens? Still no significant swelling of the rice grains. It is not until you reach near boiling temperature (212° F) that the rice begins to swell and then only slowly. Since birds cannot function at near boiling temperatures, there is no harm from eating grains of rice.

What about birdbaths and in particular, metal birdbaths? Do they present a hazard during the winter? Will a bird's feet freeze to a metal birdbath (or feeder) in subfreezing temperatures? The answer is no although the reason may not be readily apparent to you. In really cold temperatures, we risk being stuck to cold metal if we touch it with bare skin. The reason for this is that the surface of our skin is covered with many small sweat glands. Although often imperceptible, these glands cause some water loss due to evaporation at the surface of the skin. It is through this evaporative process that water is made available to instantly freeze to the surface of cold metal. Birds do not have this problem because birds do not have any sweat glands. In order to freeze to a perch, there would have to be unfrozen water available on the bird's feet or on the perch. This would not likely be the case in subfreezing temperatures.

There are a few other consider-

## Chapter 12 – Feeding Wild Birds: Facts and Misconceptions

ations to think about when attracting birds into your yard. It is best if the birds have some nearby trees or bushes or other cover to fly into. This allows them to be more sure of escape in the event of a predator. Try to place your feeder in such a way that you are not providing ready prey for neighborhood cats. There is little you can do to prevent attacks from birds of prey, especially Sharp-shinned Hawks. But this is rarely a problem and the cover of nearby bushes usually allows adequate escape. Besides, these hawks are a part of our natural world and must eat as well. If you put up nest boxes, be sure that the wood is not treated with creosote or other wood preservatives that could be harmful to birds.

Attracting birds into your yard is fun and very rewarding. With just a little forethought, we can assure our feathered friends a safe and healthy meal while they feed outside our window.

*Figure 12-2. Evening Grosbeak, another frequent winter feeder visitor in many locations.*

**Chapter 12 – Feeding Wild Birds: Facts and Misconceptions**

# Chapter 13
# Characteristics of North American Bird Families

Only the Orders, Families and Subfamilies that are represented in North America north of Mexico are listed here. Some may be on the North American list because of one or two accidental occurrences. Other families and subfamilies found in Central and South America and elsewhere in the world are not referred to here.

##  Order Anseriformes

### Family Anatidae (Swans, Geese and Ducks)

This is a large family and the birds of this family vary in size from less than one pound to over 30 pounds. (Trumpeter Swan). Only the three front toes are webbed. These birds usually have short legs and medium to long necks. Their broad, flattened bill is rounded at the end with a "nail" (unguis) or hardened, short hook at the tip of the upper mandible. They have rather dense plumage. Their powerful wings are narrow and pointed, giving strong, swift flight. Most species are gregarious and migratory. Swans are the largest of all waterfowl, followed in size by geese and then ducks.

#### Subfamily Dendrocygninae (Whistling-Ducks)
#### Subfamily Anserinae (Geese and Swans)
#### Subfamily Anatinae (Ducks)

##  Order Galliformes

### Family Cracidae (Curassows and Guans)

These medium to large-sized, somewhat chicken-like birds are found primarily in tropical forests and brushlands. They have a long tail and neck and usually have loud voices. They have strong legs but the hallux (hind toe) is at the same level as other toes of the foot (it is higher in other fowl) and they can thus grasp limbs. Much of their time is spent in trees. Their diet is mainly vegetarian and they are the only members of this order which feed their young directly. All are non-migratory. This family is represented in North America only by the Plain Chachalaca, *Ortalis vetula*.

### Family Phasianidae (Pheasants, Grouse and Quail)

These are all medium to large-sized terrestrial birds which are generally non-migratory. They have strong, powerful flight but only for short

## Chapter 13 – North American Bird Families

distances. They are heavy-bodied and have short legs with large toes for scratching and walking, and are primarily represented in North America by four subfamilies.

### Subfamily Phasianinae - (Partridges and Pheasants)

All members of this subfamily are introduced into North America. Partridges and francolins have short tails and heavy bodies, pheasants have long tails. Many of these birds have strongly contrasting patterns in the plumage and bare legs.

### Subfamily Tetraoninae - (Grouse and Ptarmigan)

Members of this subfamily are medium to large in size and have partially or fully feathered legs. Ptarmigans have feathering extending onto the toes. Short feathers cover the nostrils. These birds often exhibit elaborate courtship behaviors. Males often have colorful bare patches over or around the eyes and males of some species have unfeathered, colorful air pouches beneath the skin which are inflated during courtship displays.

### Subfamily Melagridinae - (Turkeys)

Turkeys are large bodied and the small head and long neck are naked. The head is often brightly colored with red or blue. They usually have well developed tail plumages that may be used in various displays.

### Family Odontophoridae - (New World Quail)

Quail are small, round-bodied birds with short tails. Many have crests or plumes on the head and usually a distinct plumage difference between the sexes. They are ground-dwelling birds.

# Order Gaviiformes

## Family Gaviidae (Loons)

Loons are swimming and diving birds found on both fresh and saltwater. They nest on land near shore but walk on land only with great difficulty since the legs are set so far to the rear of the body. The Red-throated Loon, *Gavia stellata*, is the only loon capable of taking off directly from the ground. All other loons must take flight from the surface of the water. The sexes are similar in appearance. Fish are the primary food source for loons but they will also take aquatic invertebrates (such as some mollusks and insects), frogs and even some plant material. The neck is thick and the bill is dagger-like. Loons have narrow, pointed wings and swift, powerful flight (speeds up to 60 m.p.h.). While in flight, the head is held lower than the line of the body. The short, strong tail has 16-20 feathers. The front three toes are webbed. Loons typically make lifelong pair-bonds and begin breeding at two to three years of age.

# Order Podicipediformes

## Family Podicipedidae (Grebes)

Grebes are not closely related to any other birds. (They were once thought to be closely related to loons.) They are swimming and diving birds found primarily on freshwater during the breeding season. Grebes may be found on saltwater during the non-breeding season. They have lobed toes (not webbed as in loons or ducks) and slender, pointed bills. Their tail consists of short, loose, non-stiff feathers. Most grebes are migratory but are weak fliers. The nest is built on the surface of the water and is usually constructed by both members of a breeding pair. The sexes have similar plumage. Grebes dive but seldom go as deep as loons. Grebes are unique in that they eat their own body feathers (this is often a chicks first meal). It is believed that these feathers help to protect the stomach from damage by fish bones.

# Order Procellariiformes
## Family Diomedeidae (Albatrosses)

This family contains the largest of the seabirds. They all have very long, pointed wings, large heads, and short necks and tails. Like all other members of this order, albatrosses have tubular nostrils, salt-excreting glands and webbed feet. In North America, they are seldom found near shore (usually 10-50 or more miles out to sea). Albatrosses can spend many months at sea. In some species, immature birds may wander across the ocean for five to seven years before returning to land to breed. When breeding does occur, each albatross female lays only a single egg. The rearing of the young takes four to five months in most species and up to one year in the largest species.

## Family Procellariidae (Shearwaters and Petrels)

Collectively, these birds are often referred to as "tubenoses" because of the tubelike nostrils (similar to albatrosses). The bills of shearwaters and petrels vary from short and heavy to long and slender and are usually hooked on the tip. The feet are webbed. They have short tails and the wings are long, narrow and pointed. Shearwaters rapidly glide just over the surface of the water. Petrels and fulmars are often more erratic in flight. As with all other members of this order, these birds are confined exclusively to ocean environments.

## Family Hydrobatidae (Storm-Petrels)

Storm-Petrels are small, sparrow to robin-sized, tube-nosed seabirds. They nest in burrows, usually on grassy offshore islands. Otherwise, their entire life is spent at sea. They have a slender bill, hooked at the tip with a single tube nostril at the base of the upper mandible. When feeding, they hover over the water and patter their feet on the surface, giving the appearance of "walking on water." The name petrel is thought to be derived from Peter and is a reference to St. Peter walking on water. The sexes are similar in plumage patterns.

Chapter 13 – North American Bird Families

#  Order Pelecaniformes

## Family Phaethontidae (Tropicbirds)

Tropicbirds are pigeon-sized seabirds which are mostly white. They have two very long central tail feathers. All four toes are webbed but the feet are set so far back that they cannot walk on land. Tropicbirds mostly nest on islands and thus avoid danger from mainland predators. They typically fly 50-100 feet over the water and plunge-dive for fish. They are migratory and wander long distances across the ocean.

## Family Sulidae (Boobies and Gannets)

Boobies and Gannets are goose-sized seabirds with very streamlined bodies and straight, sharp bills. The tail is usually long and wedge-shaped or pointed. The feet are fully webbed. They dive for fish from considerable heights.

## Family Pelecanidae (Pelicans)

Pelicans are among the largest of flying birds with wingspans of 6.5 – 9 feet. Their diet is mostly fish. Found on large, shallow, inland lakes (except the Brown Pelican which is the only truly marine species), they are very social birds and are seldom found alone. They are strong swimmers and usually feed from the surface (except the Brown Pelican which plunge-dives from heights). Pelicans may consume up to 4 lbs. of fish per day.

## Family Phalacrocoracidae (Cormorants)

Cormorants are mostly coastal, but some species are found on inland waters as well. They are duck to goose-sized, heavy bodied with a long neck, short legs and tail. The feet are fully webbed. Cormorants are mostly gregarious and generally nest colonially. They forage underwater (at depths of 80-100 feet for some species) but bring the food to the surface before swallowing it. The structure of the flight feathers is such that they are easily wettable. This allows them to compress the air from these feathers to reduce buoyancy. Contrary to popular belief, the oil gland is well developed and fully functional in cormorants and the body feathers remain waterproofed.

## Family Anhingidae (Anhingas or Darters)

Anhingas are similar to cormorants, with a long, pointed bill which is used to spear fish. The long, slender neck and head give a "snakelike" appearance (and hence, the name "Snakebird"). The feathers are wettable (even more so than cormorants). They are mainly tropical (Anhingas range from northern South America to the southern United States).

## Family Fregatidae (Frigatebirds)

Frigatebirds are large, coastal birds which spend nearly all of their time in the air. They have long, pointed wings and a long, deeply forked tail. They are extremely light (the feathers weigh more than the skeleton) and are excellent soarers. The long bill is hooked at the tip. With rather short legs, frigatebirds are very awkward on land or water and therefore are seldom found directly on the surface of either, but rather are found perched in trees or bushes when at rest. They are not migratory but often wander after the breeding season. They swoop to the ocean surface to catch fish that are near the surface and often harass other birds (gulls, boobies, etc.) until they drop the fish that they have caught. The frigatebird then catches the dropped fish in the air before it can hit the water.

# Order Ciconiiformes
## Family Ardeidae (Bitterns, Herons and Egrets)
Herons have long legs and a long neck and are usually associated with wetlands or riparian areas. They fly with the neck pulled back in a tight "S" shape while the legs trail out straight behind. Most herons have a long, pointed bill (the Boat-billed Heron, *Cochlearius cochlearius*, is a notable exception) and often have well-developed plumes during the breeding season.

## Family Threskiornithidae (Ibises and Spoonbills)
This family is made up of medium to large wading birds with elongated bodies and long necks and legs. The bill is either long and decurved or flat and spoon-shaped. The face is bare of feathers. They are usually associated with freshwater. Flight is strong and direct with the neck extended, but they seldom soar. Most are colonial nesters in trees although some nest on the ground.

### Subfamily Threskiornithinae (Ibises)
### Subfamily Plataleinae (Spoonbills)

## Family Ciconiidae (Storks)
Storks are large, heavy-bodied wading birds. Their long necks are extended straight while in flight, unlike that of herons. The rounded head has no feathers in many species, and is only partially feathered in other species. The bill is long and bluntly pointed, and slightly decurved at the tip of some species. The large wide wings enable soaring at great heights over long distances.

# Order Phoenicopteriformes
## Family Phoenicopteridae (Flamingos)
Flamingos have a rounded body, long thin legs and a very long, slender neck. The uniquely structured bill, heavy and curved at nearly a right angle in the middle, is used to strain food particles from the water assisted by the thick and elaborately structured tongue. Flamingos are highly gregarious and are strong, rapid flyers.

# Order Falconiformes
## Family Cathartidae (New World Vultures)
The members of this family are found entirely in the Western Hemisphere. They are all large, dark vultures and include the Andean Condor, *Vultur gryphus*, one of the largest flying birds (10 ft. wingspan). The head and

sometimes the neck is bare of feathers. They have a somewhat hooked bill and their talons are usually short and not sharp as in hawks. They have long, broad wings and are very adept at soaring. They may do some hissing and snapping but have no true vocalizations. They are usually somewhat long-lived (up to 50 yrs.). Most are not migratory (Black Vulture, *Coragyps atratus*, and Turkey Vulture, *Cathartes aura*, of North America are exceptions). Food is primarily carrion although some occasionally actively hunt prey (especially Black Vulture). These birds have traditionally been placed in the Order Falconiformes but much of the recent research suggests that they are much more closely related to storks and ibises.

### Family Accipitridae (Kites, Eagles, Hawks and Osprey)

(The Osprey, *Pandion haliaetus*, has sometimes been assigned its own family status, but now is generally classified as the only member of the Subfamily Pandioninae).

All members of this family have a strongly hooked, sharp bill with a membranous outgrowth (cere) at the base which is brightly colored in many species. Also noted for powerful talons and strong legs for capturing prey, they are strong flyers. The sexes are often similar in plumage patterns but females are usually larger than the males of the same species (up to twice the weight of the males in some species). They use binocular vision when hunting and have very keen eyesight, with eyes so large that in most species the ability to move them in the socket is very limited. The eyes are usually colored orange, yellow, red, or brown. Eye color is typically a function of age.

#### Subfamily Pandioninae (Osprey)
#### Subfamily Accipitrinae (Kites, Eagles and Hawks)

### Family Falconidae (Caracaras and Falcons)

A notable feature of falcons is the toothlike projection on the upper mandible near the tip (nearly lacking in Caracaras). They also have rapid, strong flight on long, pointed wings and are capable of great speeds (40-60 mph are normal speeds for Peregrine Falcons but they can attain speeds of well over 100 mph). Falcons have dark eyes and exceptionally keen eyesight. The sexes most often look alike or are very similar (American Kestrel is an exception), but the female is larger than the male.

#### Subfamily Caracarinae (Caracaras)
#### Subfamily Falconinae (Falcons)

## Order Gruiformes

### Family Rallidae (Rails, Gallinules and Coots)

Rails are small to medium-sized birds associated with water or marshes. The body appears rounded from the side but is laterally compressed. They have short tails and short, rounded wings. Rails are often secretive in behavior. Many are very poor flyers (although some species are

## Chapter 13 – North American Bird Families

migratory) and some have lost the ability to fly altogether (typical of rails on predator-free islands such as the now extinct Laysan Rail on Laysan Island). All can swim and dive but coots are the most capable swimmers with lobed toes (not webbed) similar to grebes.

### Family Aramidae (Limpkin)

The Limpkin, *Aramus guarauna*, is the only member of this family. It has a long, slightly decurved bill that is about twice the length of its head and long legs and toes. Limpkins are good swimmers but the feet are not webbed. The sexes look alike.

### Family Gruidae (Cranes)
#### Subfamily Gruinae (Typical Cranes)

Cranes are large, long-necked, long-legged birds of expansive marshlands or wet prairies and plains. Cranes hold their necks straight in front of their bodies when flying (in contrast to herons which fold their necks back). They have four sharp-clawed toes on each foot and the hallux (hind toe) is set well above the plane of the other three. The head may be partially naked or have decorative plumes on top. Most cranes usually have brown, gray or white plumage. The tails are short and they have long, wide wings. Cranes are usually migratory and often exhibit elegant courtship dances. Many mate for life. The sexes are similar but males are generally larger.

# Order Charadriiformes

### Family Burhinidae (Thick-knees and Stone-curlews)

Thick-knees get their name from the swollen leg joints that characterize these birds. Most are not found along shorelines but rather in dry fields where they feed mostly on large insects and some small animals. They are frequently active during the night. The Double-striped Thick-Knee, *Burhinus bistriatus*, is accidental in southern Texas.

### Family Charadriidae (Lapwings and Plovers)

Plovers are small to medium-sized shorebirds with thick necks, short, pigeon-like bills which are slightly swollen at the tip and large eyes. Lapwings have broad rounded wings but most plovers have long, pointed wings. Some species have slight webbing at the base of the toes (semipalmated) and the hind toe is almost completely lacking. All members of the order Charadriiformes (except Woodcocks) are diastataxic – a condition where there is a missing feather or gap in each wing between the 4th and 5th secondaries. All are strong flyers and most are migratory. Many are active both day and night.

#### Subfamily Vanellinae (Lapwings)
#### Subfamily Charadriinae (Plovers)

### Family Haematopodidae (Oystercatchers)

Oystercatchers are found along shorelines of temperate coastal waters, frequently but not always, on rocky shores. The plumage is black and white or all black and the sexes look alike. They have pink to orange feet and legs and, like other members of this order, have only three toes (no hallux), and a slight degree of webbing between the toes. The long, heavy red bill is used to remove shellfish from rocks and to open the shells. They are strong flyers with long, pointed wings and short tails.

### Family Recurvirostridae (Stilts and Avocets)

This family is represented by tall wading birds with very long legs (longer in proportion to body size than any other birds except flamingos), long, pointed wings, and long, slender bills. The bills are upturned (recurved) in avocets. The feet may be slightly webbed or nearly fully webbed. They are good swimmers and many also dive. They have short, square tails. The sexes are similar but slight differences are perceptible in some species. Both sexes share responsibility for nest building and raising of the young.

### Family Jacanidae (Jacanas)

Jacanas are short-tailed, small to medium-sized birds which look similar to gallinules. They are all tropical or semitropical. Jacanas are well known for their extremely long toes and toenails (toenails may be over four inches long in some species) which allow them to walk on the tops of floating water plants. There is a sharp, horny spur at the bend of the wing which may be used in fighting. Most are good swimmers and divers. They are non-migratory but may wander during the non-breeding season. In most species, the sexes are similar but females are usually much larger. Jacanas are among the most polyandrous of birds, and while both sexes may help with nest building, only the male incubates the eggs and raises the young. Females lay two to four clutches of eggs in the nests of two to four males, all of which she defends. She also replaces lost eggs in these nests, but otherwise plays no role in the care and development of the young.

### Family Scolopacidae (Sandpipers and Phalaropes)

#### Subfamily Scolopacinae (Sandpipers and allies)

This is a large and very diverse family. Most members have long legs and long, pointed wings. The neck varies from very short to very long and slender and the bill is generally slender. They have a small, elevated hallux (except for the Sanderling, *Calidris alba*, which lacks a hallux). Most sandpipers are highly gregarious, usually migratory and generally lack any bright plumage. They often have elaborate courtship displays. Most breed at high latitudes (many are Arctic breeders). The sexes are generally similar but plumage differences do occur in some species.

#### Subfamily Phalaropodinae (Phalaropes)

The phalaropes are the most aquatic of all shorebirds. While Wilson's Phalarope is found inland and breeds in freshwater marshes, Red and Red-necked Phalaropes are primarily oceanic

## Chapter 13 – North American Bird Families

species. These robin-sized birds have long, slender bills, long, pointed wings and long legs. The toes are lobed. They are strong swimmers. The sex roles are reversed. Phalaropes usually form monogamous pairs, but the females are more brightly colored and only the males incubate the eggs and raise the young.

### Family Glareolidae (Pratincoles and Coursers)
#### Subfamily Glareolinae (Pratincoles)
These shorebirds have long wings and a forked tail that makes them resemble a large tern. They feed mostly upon insects which they catch in the air. They may occasionally run on the ground in a manner similar to plovers. The Oriental Pratincole, *Glareola maldivarum*, is accidental in Alaska.

### Family Laridae (Skuas, Gulls, Terns and Skimmers)
Members of this family are generally medium to large-sized birds usually near or associated with water, often in marine environments. They are skillful flyers, have long, pointed wings and webbed feet, with a varied diet.

#### Subfamily Larinae (Gulls)
Gulls are typically light-colored below and light to very dark on the back as adults (there are some dark-bodied gulls and one all-white gull, the Ivory Gull, *Pagophila eburnea*,). It often takes two to three years to attain adult plumage. Immature plumages are typically brownish. The sexes usually look alike. Gulls are often thought of as coastal but many are found inland for all or part of their lives. They are good swimmers but almost never dive below the surface. There is a very large variety of items included in the diet and gulls may often play the role of scavenger. Many are migratory. They have salt glands for elimination of excess salt (as is typical of other marine birds), but they can drink salt or freshwater equally well. Most are gregarious.

#### Subfamily Sterninae (Terns)
Terns are usually smaller and more slender than gulls. Most have slender, pointed wings and rarely soar and glide as do gulls. Many terns have black on the top of the head. The sexes look alike. When feeding, they usually plunge-dive head first into the water to catch fish, which make up most of their diet. They may rest upon the surface of the water but seldom swim (they have short legs that make swimming difficult). Most are gregarious and nest in colonies.

#### Subfamily Rynchopinae (Skimmers)
Skimmers resemble large terns. They have a long bill with a heavy lower mandible that is longer than the upper mandible. A skimmer

feeds by flying along the surface of the water with it mouth open and the lower mandible "skimming" the water. When the lower mandible strikes food, the mouth snaps shut. They have a forked tail and very short legs. The sexes are alike but females are usually smaller.

**Family Stercorariinae (Skuas and Jaegers)**
Members of this subfamily are generally the largest of the entire family. They have short, strong legs and a bill that is usually somewhat hooked at the tip with a fleshy cere at the base of the bill around the nostrils. In jaegers, the center two tail feathers are longer than the rest of the tail. All of these birds are predatory and hunt or frequently chase other birds (notably gulls) until the chased bird drops the fish it has caught or disgorges the fish it has just eaten.

**Family Alcidae (Auks, Murres and Puffins)**
This family is strictly marine. Alcids are small to medium sized, heavy-bodied birds. They are frequently found along coastal cliffs. They have short, narrow wings and a somewhat labored flight. They use their wings to "fly" underwater. Alcids have short tails and the plumage is usually black or black and white. Most nest on offshore islands or on coastal cliffs. (The Marbled Murrelet, *Brachyramphus marmoratus*, is an exception which nests in the treetops of old-growth coastal forests as much as 50 miles inland.) The sexes look alike. Many spend the winter entirely out of sight of land, thus winter habits and movements are poorly understood for most species.

# Order Columbiformes

**Family Columbidae (Pigeons and Doves)**
Members of this family typically have a small head on a short neck. The bill is short with a slit-like nostril. They have soft, dense plumage. The sexes usually look alike. Some species lack an oil gland. The feathers are strong but loosely attached and can be dislodged easily. They are generally strong flyers. There are 11 primary feathers on each wing but only 10 are functional. The legs are usually short with four strong toes. Species in this family are the only North American birds capable of drinking water by sucking it up (most birds dip the bill, then tip the head back to drink). They have the most specialized crop of any bird, producing "pigeons milk" during the breeding season. This "milk," produced by both parents, is rich in fat, lecithin and protein (no sugar). This is the only meal provided to the young for three to five days, but the "milk" is produced and fed to the chicks for up to 18 days. There is no real distinction between doves and pigeons. Dove is usually used to denote smaller birds but the two terms are interchangeable.

# Order Psittaciformes

**Family Psittacidae (Parrots, Macaws, Lories and Parakeets)**
These are primarily forest dwelling birds (predominately tropical). They

eat some insects and nectar but mostly eat nuts, seeds and fruits. They have zygodactyl feet (1st and 4th toes point backward, 2nd and 3rd toes point forward) and often hold food in one foot while eating. Most species are usually very gregarious outside of the breeding season. The sexes are alike. They are non-migratory. Species in this family are characterized by having a short, heavy, large bill with the top mandible hinged and sharply decurved. This beak is used to aid climbing. They have a very direct flight on rapidly beating, stiff wings. The Thick-billed Parrot, *Rhynchopsitta pachyrhyncha*, (now rare even in Mexico) occasionally ranged into Arizona in the early part of the 1900s. The Carolina Parakeet became extinct in 1917. All other members of this family are introduced into North America north of Mexico.

### Subfamily Platycercinae (Australian Parakeets & Rosellas)

The Budgerigar, *Melopsittacus undulatus*, is the only member of this subfamily in North America that has established feral, but declining population in Florida.

### Subfamily Psittacinae (Typical Parrots)

The Rose-ringed Parakeet, *Psittacula krameri*, is the only member of this subfamily in North America that has established feral populations in parts of California and Florida.

### Subfamily Arinae (New World Parakeets, Macaws, & Parrots)

# Order Cuculiformes

## Family Cuculidae (Cuckoos, Roadrunners and Anis)

Cuckoos are medium to large-sized birds with zygodactyl feet. Most occur in forested or dense brushy areas. Many species are well known as brood parasites (none of the species in North America are parasitic). They have a long bill and often have a long tail.

### Subfamily Cuculinae (Cuckoos)

### Subfamily Neomorphinae (Ground-Cuckoos, Roadrunners)

### Subfamily Crotophaginae (Anis)

# Order Strigiformes

## Family Tytonidae (Barn Owls)

Barn Owls are similar to typical owls. They have a heart-shaped facial disk with eyes that are dark and usually smaller than those of typical owls. Their long, slender legs are usually feathered. Barn Owls are light-colored, especially from below and they have a short, square tail. During the molt, the tail feathers are replaced from center outward (the reverse is true of typical owls). The middle toe of each foot is pectinate (has small, toothlike projections) along the inner edge. Predominately non-migratory, many can locate prey in total darkness, and they probably mate for life.

### Family Strigidae (Typical Owls)
Owls are found worldwide except for Antarctica and some oceanic islands. They have very keen eyesight with large eyes that are often fixed or have very limited movement (compensated for by extreme ability to turn the neck). Owls have a well-developed nictitating membrane and well-developed hearing with ears often quite asymmetrical in structure. Most are crepuscular or nocturnal but some species hunt during the daylight hours. The legs are feathered in most owls. Owls have four toes with the outer toe of each foot reversible to point forward or backward and they have strong talons on each toe. The strongly hooked bill has a cere at its base. Night hunting owls fly silently. All owls regurgitate pellets of undigested matter (bone and fur). The sexes are similar but the female is usually larger.

## Order Caprimulgiformes
### Family Caprimulgidae (Goatsuckers, Nighthawks and Nightjars)
These birds are largely nocturnal or crepuscular insectivores (although some are active during a part of the day). The beak is small but the mouth is very large with rictal bristles on each side which aid in catching insects. They usually have very distinctive vocalizations. Most have a long tail. The eggs are laid directly on the ground and little or no nesting material is used. The sexes are usually similar but with some subtle differences between male and female. They have long, pointed wings and feet that are small and weak. They do not sit crosswise on a branch as most birds do but rather sit lengthwise along the branch. The foot has a "feather comb" (the pectinate margin of the middle toe of each foot). They have large, dark eyes. The soft plumage is usually cryptically colored brown or gray. Some species are migratory. The Common Poorwill, *Phalaenoptilus nuttallii*, of North America is the only bird known to exhibit true hibernation (although many species of unrelated birds can go into torpor for short periods or up to a few days).

#### Subfamily Chordeilinae (Nighthawks)
#### Subfamily Caprimulginae (Nightjars)

## Order Apodiformes
### Family Apodidae (Swifts)
Swifts spend most of their waking hours in the air. Feeding and mating takes place in the air (some copulatory activity also occurs at the nest site). At least one species (Common Swift, *Apus apus*, of Eurasia) is known to sleep in the air. They have slender, tapered bodies and long, pointed wings and many have a short, stiff tail of 10 feathers. Much of the wing surface is made up of the 10 long primaries (the fleshy part of the wing and the wrist is short and close to the body). The wings are moved in rapid stiff wing beats. The feet are strong and used to perch on vertical

## Chapter 13 – North American Bird Families

surfaces but the legs are very short and weak. Swifts cannot perch on a wire or branch as swallows do, and if forced onto the ground, may have much difficulty getting back into the air because their weak legs cannot help push them away from the surface. The sexes look alike. They have a large mouth with a short, slightly decurved bill. They are very gregarious. The flight is very rapid, often close to 100 mph. Bursts of speed of approximately 200 mph have been reported for some species.

### Subfamily Cypseloidinae (Cypseloidine Swifts)
### Subfamily Chaeturinae (Chaeturine Swifts)
### Subfamily Apodinae (Apodine Swifts)
## Family Trochilidae (Hummingbirds)
### Subfamily Trochilinae (Typical Hummingbirds)

This is the largest nonpasserine bird family and the second largest family of birds in the Western Hemisphere. This family contains the smallest of all birds and all are found only in the Western Hemisphere. Hummingbirds have the unique ability to hover, can fly backwards equally well as forward and can also fly up and down. They are mostly tropical (although the Rufous Hummingbird, *Selasphorus rufus*, summers as far north as Alaska). They have a long, needlelike bill and the tongue usually extends well beyond the length of the bill. Their long, pointed wings have long primaries and relatively short secondaries. Hummingbirds feed mainly on nectar, insects and spiders. Much of the plumage is iridescent in both sexes but the more extensive and brighter patches occur in the male. Males have complex courtship displays but frequently pay little or no role in caring for the young. The high metabolic rate often requires a torpid condition at night to prevent starvation.

# Order Trogoniformes
## Family Trogonidae (Trogons)
### Subfamily Trogonidae (Trogons)

Trogons are found in pantropic forests and woodlands. They have a short, broad bill (with serrated edges in some species) and a long tail. There are usually strikingly contrasting patterns of coloration. The sexes often look alike although males are brighter in some species. They often exhibit bright, metallic colors (frequently green) over the back and center tail feathers.

# Order Upupiformes
## Family Upupidae (Hoopoes)

Hoopoes are very distinct Old World birds that are not closely related to any others. They have a long bill and a long crest on the head that may be raised like a fan. These cavity-nesting birds generally probe into the soil to feed upon insects. There is a single record of an Eurasian Hoopoe, *Upupa epops*, occurring in Alaska.

# Order Coraciiformes
## Family Alcedinidae (Kingfishers)
### Subfamily Cerylinae (Typical Kingfishers)
Kingfishers measure from 4-18 inches in length. They are most often associated with water and many are specialized for catching fish. (Not all kingfishers eat fish. Some eat large insects.) Most kingfishers are heavy-bodied with a short tail (a few species have longer tails). They have a short neck and a large-appearing head. The beak is long, pointed and dagger-like. The short legs have feet that are rather weak with the 3rd and 4th toes joined together and the 1st toe (hallux) pointing backwards. The sexes look alike or very similar. Kingfishers are usually found singly (except for breeding pairs) and most feed on fish which are caught by diving head first into the water. The fish-eating kingfishers all dig burrows along the banks of streams and rivers for nesting.

# Order Piciformes
## Family Picidae (Woodpeckers and allies)
### Subfamily Jynginae (Wrynecks)
The Eurasian Wryneck, *Jynx torquilla*, is accidental in Alaska.
### Subfamily Picinae (Woodpeckers)
Woodpeckers are adapted for clinging and hammering on trees. The bill is long and sturdy with the upper mandible being chisel-like. The short, stiff tail is used as a prop when the bird is perched on a tree trunk. Woodpeckers have a thick skull and strong neck muscles. Most have very long tongues (sapsuckers have much shorter tongues than most other woodpeckers). The feet are zygodactyl (the hallux is missing from all three-toed woodpeckers).

# Order Passeriformes
## Family Tyrannidae (Tyrant Flycatchers)
Tyrant Flycatchers are a diverse group. Most have a broad, flattened bill which is slightly hooked at the tip. Bristles around the base of the bill help in the capture of insects as the mouth is opened. Flycatchers generally have an upright posture. There are 10 primaries per wing. Insects are the main diet but some eat berries or, occasionally, small fish. All North American species are migratory. They have short and weak feet and legs.

## Chapter 13 – North American Bird Families

### Subfamily Elaeniinae (Tyrannulets, Elaenias and Allies)
### Subfamily Fluvicolinae (Fluvicoline Flycatchers)
### Subfamily Tyranninae (Tyrannine Flycatchers)
## Family Laniidae (Shrikes)

Shrikes are the only truly raptorial songbirds. They have a grayish plumage with a black facial mask and black in the wings and tail. The strong bill is hooked with a toothlike projection near the tip of the upper mandible which fits in an opposing notch on the lower mandible. They are solitary birds of open country and feed on small rodents, large insects and small birds. They impale uneaten prey on thorns, sharp twigs or even barbed wire and return to feed on it later (this has earned them the nickname "butcherbird"). Both North American species are migratory.

## Family Vireonidae (Vireos)

Vireos are small to medium-sized birds of the Western Hemisphere. Most have dull, grayish green plumage (vireo is Latin for "green bird"). They are shy and sluggish in behavior. The bill is short and straight and the upper mandible has a small hook at the tip. The nest is a woven cup hanging in the fork of a small branch. The sexes are similar. Bristle-like feathers partially cover the forehead and nostrils. There are 10 primaries per wing but the outermost is very poorly developed.

## Family Corvidae (Jays, Magpies, Crows and Ravens)

Corvids are medium to large-sized birds found in a wide variety of habitats nearly worldwide. They are often quite vocal and aggressive. They have a sturdy bill, often sharply tapered. The sexes look similar. Corvids are often considered to be among the most intelligent of birds and often show complex social behaviors. There are 10 primary feathers on each wing. The nostrils are covered by small, dense, stiff feathers (lacking in the Pinyon Jay, *Gymnorhinus cyanocephalus,*). There are rictal bristles at the base of the bill. Corvids are strong flyers and generally non-migratory. Some of the larger members of the family are long-lived and some mate for life. The diet is extremely varied and general.

## Family Alaudidae (Larks)

This is mainly an Old World family represented in North America only by the Horned lark and the introduced Sky Lark, *Alauda arvensis*. None are found on oceanic islands. They are open country, ground-dwelling birds. Most have elaborate courtship flight displays with complex songs. The sexes mostly look alike. The hind claw is straight and unusually long. Larks walk instead of hop. They are typically gregarious but not colonial nesters.

## Family Hirundinidae (Swallows)
### Subfamily Hirundininae (Typical Swallows)

Swallows are aerial insectivores often mistaken for swifts but not at all related to swifts. They have short necks, long, pointed wings and small, broad, flattened bills with large mouths. The sexes are often similar. They are gregarious and many are colonial nesters. There are 12 feathers in the tail (swifts have 10) with the outer feathers longer than the rest of the tail (sometimes very much so, as in the Barn Swallow, *Hirundo rustica*,). Swallows spend more time in the air than most other songbirds. The plumage is often iridescent and usually darker above and white or lighter below. They can perch (unlike swifts) but walk on the ground only with much difficulty.

### Family Paridae (Chickadees and Titmice)

This family of small (usually less than 6 in. long), round-bodied birds has soft, thick plumage which is usually a soft gray to brown in color. The small, stout bill has nostrils covered or partially covered with bristles (as in crows and jays). Most have a long tail. They have long legs with small but strong feet. The rounded wings have 10 primary feathers each. The 10th primary (outermost) is only half the length of the 9th primary. They are always very active and curious and often rather tame around people. Many are migratory to some extent. The sexes look alike. All are cavity nesting.

### Family Remizidae (Penduline Tits and Verdins)

These birds are similar to titmice but construct tightly woven, pendulant nests. They are mostly non-migratory.

### Family Aegithalidae (Long-tailed Tits and Bushtits)

These tiny birds are similar to titmice. They typically have very long tails and are weak flyers but are constantly on the move through the underbrush. They are highly social.

### Family Sittidae (Nuthatches)

#### Subfamily Sittinae (Nuthatches)

Nuthatches are small, round-bodied birds with a large head, short neck and a tail which is short and may be rounded or squared. They have short legs and feet with four long toes. Nuthatches are noted for their ability to walk head first down a tree trunk as easily as going up. They are the only birds which routinely forage in this manner. Their long, slender bill has an upper mandible which is chisel-shaped at the tip. The lower mandible tapers slightly upward near the tip. The nostrils are partially covered with stiff bristles. Their call is short and nasal. The sexes are similar but males are sometimes a bit more intensely colored. There are 10 primaries per wing. Nuthatches are mostly non-migratory and all are cavity nesting.

### Family Certhiidae (Creepers)

#### Subfamily Certhiinae (Creepers)

## Chapter 13 – North American Bird Families

Creepers are found in the Northern Hemisphere with most species occurring in Eurasia. The Brown Creeper, *Certhia americana*, is an exception as its range extends south to Nicaragua. It is the only member of this family that is found in the New World. Creepers are small, tree dwelling birds with long, stiff tails that are used to prop themselves against the tree trunk (like woodpeckers). They have a slender, decurved bill that is used for probing into crevices in the bark. The coloration is usually gray or light below and brown to black above. The toes are long with long, sharp claws. Males and females look alike.

### Family Troglodytidae (Wrens)

Wrens are small (except for the Cactus Wren, *Campylorhynchus brunneicapillus*, of North America), brownish birds of the Western Hemisphere (except the Winter Wren which is also found in Europe). They have slender, slightly decurved bills. Many have a long tail which may often be held cocked up over the back. Wrens have long legs and toes. Usually the male builds the nest and marks the territory by building many "false nests" around the perimeter of the territory. Most wrens have very musical songs. The sexes are similar. They are very active but flight is usually rather weak. Wrens are mostly solitary and most are non-migratory except for a few northern species.

### Family Cinclidae (Dippers)

Dippers are the only aquatic passerines. They are found only along streams and waterfalls, and actively forage around rocks at the bottom of swift moving streams. They have a conspicuous nictitating membrane that covers the eye as they enter the water and a large preen gland which my aid in waterproofing the feathers. (This gland is about 10 times larger than in most song birds). The short tail is usually cocked in a manner similar to that of wrens. The sexes look alike and the birds are solitary except during breeding. Weak pair-bonds are made and males typically do not remain at the nest site or play a strong role in care of the young. Dippers use their wings to "fly" underwater. Their rather musical songs are sung throughout the entire year.

### Family Pycnonotidae (Bulbuls)

This large, Old World family is represented in North America only by the introduced Red-whiskered Bulbul, *Pycnonotus jocosus*, in southeast Florida. They have slender bills (frequently decurved to some degree), long tails and short, rounded wings. The sexes look similar. Bulbuls are primarily forest birds but many have adapted to agricultural environments. The soft plumage is rather drab in most species. On the nape is a patch of filoplumes (thin hair-like feathers which lack a vane) which are more densely packed than that found on the nape of most birds. Bulbuls are gregarious and often quite vocal.

### Family Regulidae (Kinglets)

Kinglets are among the smallest of all birds except hummingbirds. They are usu-

### Chapter 13 – North American Bird Families

ally very active and often flit their wings. They have 10 primaries per wing.

### Family Sylviidae (Old World Warblers and Gnatcatchers)

These birds are related to Old World flycatchers and thrushes, and are not so closely related to American wood warblers. The plumage is generally somewhat drab. There are 10 functional primaries per wing (American wood-warblers have only nine).

#### Subfamily Sylviinae (Old World Warblers)
#### Subfamily Polioptilinae (Gnatcatchers and Gnatwrens)

### Family Musicapidae (Old World Flycatchers and allies)

These birds have broad, flattened bills with conspicuous rictal bristles at the base of the bill (as in the Tyrant Flycatchers) which aid in catching insects. They have short legs. They are mostly accidental in the New World but three species have extended their range and now have sporadic distribution in western Alaska.

### Family Turdidae (Solitaires, Thrushes and allies)

This family is represented nearly worldwide. Many have complex, very musical songs. The tarsus (long part of the foot between the ankle and toes) is "booted" (smooth, unscaled). Most species show spotting on the breast at some stage in their lives (for some, such as American Robin, *Turdus migratorius*, this occurs only in the juvenal plumage). There are 10 primaries per wing.

### Family Timaliidae (Babblers)

Babblers have plumage that is generally soft, often subdued and long and thick on the lower back. The legs and feet are strong. There is much variability in other characteristics: tails are short to very long, the bills are small to long and sickle-shaped or massive and laterally compressed. The bills are hooked at the tip in many species. The wings are usually short and rounded. This family is represented in North America only by the Wrentit, *Chamaea fasciata*, of Oregon and California.

### Family Mimidae (Mockingbirds, Thrashers and allies)

This New World family is made up of medium-sized birds, most of which have long tails (often cocked up), short, rounded wings with 10 primaries each and bills that are slightly pointed and of medium to long length. They have strong legs that are usually long. Many are excellent singers and often imitate other bird songs (the Northern Mockingbird, *Mimus polyglottos*, is especially noted for this). They usually have long rictal bristles. The sexes look alike. These birds are often found near or on the ground.

### Family Sturnidae (Starlings and allies)

This is an Old World family – European Starling, *Sturnus vulgaris*, Hill Myna, *Gracula religiosa*, and Crested Myna, *Acridotheres cristatellus*, are introduced to North America. Round-bodied birds of medium size, with short, square tails and strong legs, most have glossy, iridescent plumage. Short, pointed wings have 10 primaries each and the flight is strong and direct. Sexes are similar. Very gregarious, they often are skilled at mimicking other bird songs and calls (even human voices). Most walk rather than hop.

### Family Prunellidae (Accentors)

This is an Old World family (the Siberian Accentor, *Prunella montanella*, has strayed into North America on a few occasions). Accentors have thin bills and are rather secretive, preferring to spend most of their time on the ground. Insects make up the bulk of the diet, although berries and some other plant material become more important during the winter as the insect population decreases.

### Family Motacillidae (Wagtails and Pipits)

Birds in this family are small (typically sparrow-sized) and slender with slender, pointed bills. There are only nine functional primaries per wing and they are generally strong flyers. The hind claw is long (but not as long as in larks) and all have long tails. Many have white outer tail feathers. The tail is especially long in wagtails, which pump the tail up and down constantly. They frequently call and sing while in flight. Most species are migratory.

### Family Bombycillidae (Waxwings)

Waxwings are small songbirds of the Northern Hemisphere. They have soft plumage, crests on the back of the head and a black facial "mask." The short bill has a broad base and is notched and hooked at the tip. There are 10 primaries per wing and some of the secondaries (and to a lesser extent the tail) have elongated shafts, some of which end in a bright red drop-shaped waxy material. The sexes look alike. Waxwing flocks often wander greatly, especially during the winter.

### Family Ptilogonatidae (Silky-flycatchers)

This family is found from Central America to the Southwestern United States in dry, brushy habitats. They have soft, silky plumage, a crest on the head and a long tail. They have short, broad bills with rictal bristles at the base. They are very gregarious (except for one species). The sexes are dissimilar. They have somewhat erratic wing beats while in flight. Phainopepla, *Phainopepla nitens*, is the only member of this family found in the United States.

### Family Peucedramidae (Olive Warbler)

Like the wood-warblers, with which it has traditionally been grouped, the Olive Warbler, *Pheucedramus taeniatus* has nine primaries per wing. Recent studies of DNA hybridization data as well as a better understanding of morphology and breeding biology of this bird now suggest that the Olive Warbler should belong to its own family. It is primarily insect-eating and shares many of the characteristics that are found in the wood-warblers.

## Chapter 13 – North American Bird Families

### Family Parulidae (Wood-Warblers)
This is a New World family of small, active insectivores with straight, thin bills and long, slender legs and toes. The plumage is often very bright with sexes generally being similar in the tropics and subtropics. In species of temperate climates, males mostly have the bright breeding plumage and females usually have a much duller plumage. The majority of temperate species are migratory. There are nine primaries per wing (Old World Warblers have 10).

### Family Coerebidae (Bananaquit)
The Bananaquit, *Coereba flaveola*, is the only member of this family. It is a small and active bird. Its bill is curved and sharply pointed. The plumage is brightly colored. It was formerly grouped with honeycreepers.

### Family Thraupidae (Tanagers)
Tanagers are found only in the New World and most species are tropical. They are frequently found high in the trees and are often secretive in behavior. They generally have bright plumage (some of the more northerly species have dull juvenile and/or female plumages). Most are usually monogamous and remain paired throughout the year. There are nine primaries per wing. They have rictal bristles around the base of their stout bills. The song is weak or even lacking in many species. North American tanagers (north of Mexico) have recently (2009) been moved into the Family Cardinalidae.

### Family Emberizidae (Emberizids)
This large family of birds has short, conical bills (not as large as Family Cardinalidae). There are nine primaries per wing and 12 tail feathers.

### Family Cardinalidae (Cardinals, Grosbeaks, N. Am. Tanagers and allies)
This family has sexes which are dissimilar. The male is often brightly colored. They have a short, conical bill and many are excellent singers (even the females of some species may sing). They are often gregarious except when nesting. Most temperate species are migratory. There are 9 primaries per wing and 12 tail feathers.

### Family Icteridae (New World Blackbirds and Orioles)
This is a family that exhibits much diversity. The plumage varies from glossy black to very brightly colored. The sexes are generally dissimilar and the males of many species are polygamous. There are nine primaries per wing. Many nontropical species are migratory.

### Family Fringillidae (Finches)
These finches are small to medium-sized birds with short, conical bills which may be quite large in some species (such as the Evening Grosbeak, *Coccothraustes vespertinus*). The sexes are dissimilar with immature males looking much like females. Some give calls or songs while in flight. There are nine primaries per wing. Many species in this group feed regurgitated,

## Chapter 13 – North American Bird Families

whole seeds to their newly hatched young (unlike most seed-eating birds which usually feed insects to their young for the first several days).

### Subfamily Fringillinae (Fringilline Finches)
### Subfamily Carduelinae (Cardueline Finches)
### Family Passeridae (Old World Sparrows)

This family is Old World in distribution (House Sparrow, *Passer domesticus*, and Eurasian Tree Sparrow, *P. montanus*, are introduced into North America). They are very gregarious. The sexes may or may not be similar. They have short, finch-like bills, short legs, 10 primaries per wing and 12 tail feathers. They are mostly non-migratory.

### Family Ploceidae (Weavers)

These are small to medium-sized birds with conical bills and short tails although, males of a few species develop long, ornate tails during the breeding season. They are mostly sedentary and usually nest colonially. Their bright plumage, especially in the males, often makes them popular as cage birds. Most species are found in Africa but a few species are found in Eurasia and Australia. This family is represented in North America by the Orange Bishop, *Euplectes franciscanus*, which as a small population in Phoenix, Arizona and a small, declining population around Los Angles.

### Family Estrildidae (Estrildid Finches)

This is a large family from the Old World tropics, mostly Africa. These are small or tiny birds with bright plumage that varies little from season to season. The large conical bill is often brightly colored. Primarily seed-eating birds that are highly gregarious in the non-breeding season but usually breed in solitary pairs. Many are popular cage birds. In North America, the Nutmeg Mannikin, *Lonchura punctulata*, has established feral populations in parts of southern California.

# Chapter 13 – North American Bird Families

# Dictionary of Ornithological Terms

## — A —

### Accipiter
Genus name for the "true" or "short-winged" hawks. In North America, these are Sharp-shinned Hawk, Cooper's Hawk and Northern Goshawk.

### Adductor muscles
The muscles which close the jaws. Strong and powerful in hawks, eagles, owls and many seed-eating birds but rather weak in most insect-eating birds.

### Aerie
A bird's nest. Most often refers to a large nest of a bird of prey, especially cliff-dwelling birds.

### Afterfeather
A second feather, often small, branching from the inner main base of a body feather. In emus and cassowaries, this feather is nearly the same length as the main feather.

### Aftershaft
The shaft of an afterfeather (see above).

### Albumen
The protein and membrane layers which make up the white of an egg.

### Alcid
General term used for a member of the Auk Family.

### Allopatric species
Closely related species that are isolated geographically.

### Allopreening
Behavior in which one bird preens the feathers of another individual, often a mate. Also called "mutual preening."

### Altricial
Term referring to birds whose young hatch in helpless or near helpless conditions. Many are hatched nearly naked and with their eyes closed.

### Alula
A small, feathered projection just beyond the wrist of a bird's wing. Acts as a wing slot to help direct air over the upper surface of the wing and reduce turbulence and stalling. Also called the "bastard wing."

### Anisodactyl
The arrangement of toes on a bird's foot wherein three toes point forward and the first toe (the hallux) points backward. This is the most typical foot pattern in birds and is found in all Passeriformes (songbirds).

### Antiphonal singing
Singing pattern where both sexes of a mated pair alternate singing. Each may sing the same song or a different song, depending upon the species. Also called "responsive singing."

### Apteria
The regions of bare skin between the feather tracts, occasionally with some down or semiplume feathers.

### Auriculars
"Ear coverts" or feathers covering external ear opening on a bird's head.

## Axillars

The innermost feathers lining a bird's wing, form the armpit or wing pit area between the wing and body. Also called axillaries.

## — B —

## Barb

The primary and largest branches along the shaft of a feather (beyond the region where any afterfeather may be found). These barbs (and their associated barbules and barbicels) compose the vane of the feather.

## Barbicel

Microscopic hooks on the barbules of a feather which lock it to barbules from an adjacent barb; help keep the vanes of the feathers from becoming separated.

## Barbule

Small, numerous branches arising from each side of a feather's barbs.

## Bastard wing

See alula.

## Billing

Behavior where mated pairs touch or clasp each other's bills, frequently a pattern of courtship.

## Biochromes

Naturally occurring chemical pigments which cause the structural colors observed on a bird.

## Bolus

A compacted ball or mass of food swallowed by a bird. Sometimes, as with the Gray Jay, these sticky food masses may be stored or hidden so that the bird may feed upon them later.

## Bristle

A short, stiff feather almost totally lacking in barbs. Found almost exclusively on the head, bristles provide protection and some sensory input.

## Brood

The number of young birds that hatch from a clutch of eggs.

## Brooding

An adult bird using its body to provide warmth and cover for its group of young (brood). Some birds are brooded in the nest and others are brooded outside of the nest site.

## Brood parasite

A bird which lays its eggs in another bird's nest. This may be an occasional or a very regular act. Birds such as the Brown-headed Cowbird are called *obligate parasites* because they never build a nest or raise their own young and must always lay their eggs in the nest of other species.

## Brood patch

Bare, loose patches of skin on the breast or abdomen of a bird by which heat is transferred to incubating eggs.

## — C —

## Calamus

The portion of a feather's shaft extending from the point at which the vane begins to where the feather is attached to the skin.

## Carina

See Keel.

## Caudal

Referring to the tail or tail region; in the area of the tail.

## Cere

The thickened, membranous covering over the base of the maxilla, especially in birds of prey.

## Cerophagy

The eating of wax. In birds, refers to birds which eat and digest wax. Many birds may eat berries with waxy coatings, but only a few birds (such as Yellow-rumped Warblers and Tree Swallows) can actually digest the wax.

## Chalaza

Thick, spiraled portion of albumen in an egg that helps to keep the yolk properly suspended within the egg.

## Cloaca

A common passageway in birds at the end of the urogenital and digestive tracts. Sperm or eggs, feces and urine are all expelled from the body by the cloaca.

## Cloacal kiss

In most male birds there is no structure analogous to a mammalian penis. During mating and copulation, the cloaca of the male and female briefly come in contact. It is during this "cloacal kiss" that sperm is transferred from the male to the female.

## Clutch

All of the eggs of one female bird that are laid during a single nesting period. Some birds lay two or more clutches during the entire breeding season.

## Cob

A male swan.

## Commensalism

Specific behaviors or an entire lifestyle which benefits from habits of another species without causing that species any harm. For example, Cattle Egrets follow cattle (water buffalo in their native Africa) and gather insects stirred up by the hooves of the cattle. Cowbirds earned their name for this same behavior with American bison.

## Congeneric

Refers to birds (or any organisms) which belong to the same genus.

## Conspecific

Term for individuals or groups of individuals belonging to the same species.

## Contour feather

The vaned outer feathers of a bird's body which form the outline or "contour" of the bird's shape.

## Corniplume

A tuft of erect feathers on a bird's head such as the "ears" on a Great Horned Owl or the small tufts above each eye of a Horned Lark. These tufts are paired, as opposed to a crest which is a single projection of feathers.

## Coverts

The usually short feathers covering the upper surface of the fleshy portion of the wing and over and below the tail. The upper tail coverts in the male Common Peafowl have become exceedingly large and ornamental and are erected and fanned out during display. Such feathers are often incorrectly referred to as the tail, which is really much smaller and more inconspicuous.

## Covey

Refers to a group of partridges, quail or other game birds.

## Crèche

Nursery or nesting area where young seabirds are reared. Often used to refer to a group of penguin chicks.

## Crepuscular
Most active during the twilight hours of dawn and dusk.

## Crest
A single, but prominent tuft of feathers on the head of a bird. Frequently held erect or at least may be erected. California Quail and Steller's Jay are examples of birds with crests.

## Crissum
The undertail coverts in birds where feathers are differently colored from the rest of the bird's undersides. One example is the deep rusty undertail coverts of the Bohemian Waxwing.

## Crop
An enlargement of the esophagus in the region of the neck used primarily for temporary storage of food. Not all birds have crops and of those that do, the size or degree of development is widely variable from species to species. Many insect-eating birds have little or no crop.

## Crop milk
A milky, protein-rich secretion of the crop that is fed to the young during the first few days after hatching. This is best known in pigeons and doves but occurs in a few other species as well (sandgrouse and some tubenoses).

## Crown
Region on top of a bird's head between the forehead and back of the head.

## Culmen
Ridge along the upper portion (maxilla) of a bird's beak.

## Cursorial
A way of life adapted to walking or running such as an Ostrich or Emu.

## Cygnet
The young chick of a swan.

# — D —

## Decurved
Curved downward. Usually refers to the shape of a downward curving bill such as on a curlew or Brown Creeper.

## Depressor muscle
The large muscle mass overlaying the sternum which pulls the wings down and forward in the downbeat or power stroke.

## Determinate egg-laying
Term used for birds which have a fixed number of eggs laid per clutch.

## Dewlap
The fleshy, unfeathered and often brightly colored pendant skin that hangs from the lower portion of the bill of some birds, such as turkeys. (See also, Lappet and Wattle.)

## Diastataxic
A condition wherein there is a large gap between the 4th and 5th secondary feathers on the wings. All shorebirds (except woodcock), gulls and terns as well as many other non-passerine birds have this condition. All passerine birds are *eutaxic*, that is, they lack this gap.

## Didactyl
Having only two toes on each foot, as does the ostrich.

## Distal
In the direction of the farthest point from the point of reference. Opposite of *proximal*.

## Ornithological Dictionary

### Diurnal
Term used to describe birds (and other animals) that are active during the daylight hours.

### Diverticulum
A "false crop" or expansion of the esophagus which is "diverted" from the main esophagus and used as a food storage organ. Found especially in redpolls and crossbills.

### Dorsal
Referring to the back or upper surface.

### Dueting
Simultaneous singing of the same song by both members of a mated pair of birds. Usually the same (or nearly the same) song but it may also be a complex song made up of two components, one sung by the male and one sung by the female.

### Dummy nest
A nest, often incomplete, built by the male. Its function is not fully understood, but may help define the territory. In North America, this behavior is most notable in House Wren, Marsh Wren and Prothonotary Warbler.

### Dump nest
Nest, often poorly constructed and incomplete, that contains eggs from two or more hens, mostly ducks and pheasants. These eggs are rarely (if ever) incubated.

## — E —

### Ecdysis
In birds, the act of molting.

### Eclipse plumage
The dull, female-like plumage of male ducks occurring during the late summer. This more cryptic plumage coincides with the time that the males are flightless due to molting all of their primaries (flight feathers), increasing their need for more protective coloration.

### Ectoparasite
Any parasite which lives on the outside of the body.

### Egg tooth
A small, horny protuberance on the tip of the upper half of the beak on a bird at the time of hatching. Used to help crack the egg shell from within just before hatching.

### Elevated
A reference to the position of the hallux (hind toe) on birds where the hallux is raised above the ground.

### Endoparasite
A parasite which lives within the body of its host.

### Erythrism
Condition where feathers have excess of reddish-brown pigment. Not common in birds but accounts for some different color phases in some species.

### Eutaxic
The opposite of diastataxic. Without the space found between the 4th and 5th secondaries of diastataxic birds. All birds in the order Passeriformes, are eutaxic as are many other species.

## — F —

### Falcate (or falciform)
Having the shape of a sickle; hooked or curved.

## Ornithological Dictionary

### False crop
(see Diverticulum)

### Feather comb
For some species of birds, the edge of the middle claw has tiny "teeth" (called pectinations), that may be used to help remove parasites and in other types of feather maintenance. Most noted in nightjars, which have a feather comb on the inner edge of the middle claw, and herons, where it is on the outer edge of the middle claw. Also found in Barn Owls, herons and bitterns.

### Fecal sac
The membrane-bound excrement or droppings of young nestlings, particularly songbirds.

### Femur
The large bone of the upper leg.

### Filoplume
A specialized, nearly barbless feather which grows near the base of a contour feather. Sparsely distributed and found mainly on the nape of the neck and on the wings. Movement of these filoplumes provides sensory input about such details as airspeed.

### Fledgling
A young bird out of the nest but still under parental care.

### Frugivorous
Diet that consists mostly of fruits.

### Furcula
A bird's "wishbone:" the two clavicles which have become fused together.

### Fusiform
Having a shape that is tapered at both ends or "spindle-like."

## — G —

### Gander
A male goose.

### Gape
The opening between the upper and lower parts of the bill when the bird has its mouth open.

### Gaping
Behavior performed by altricial nestlings, of opening the mouth as a way to solicit food from the parents.

### Gizzard
The muscular portion of a bird's stomach. Serves the same role as the grinding teeth of mammals.

### Gorget
Iridescent feathers (collectively) on the throat of a male hummingbird.

### Graduated tail
A bird's tail in which the central feathers are the longest and each feather successively outward from the center is markedly shorter.

### Gramnivorous
Having a diet of grass. Many geese and ducks are largely gramnivorous.

### Granivorous
Having a diet mainly of seeds.

### Gular pouch
A somewhat expandable region of the upper throat. The pouch of pelicans and cormorants, for example. Some birds do gular fluttering during panting as a method to help with cooling.

### Gynandromorph
A genetic mixture of male and female characteristics in one individual.

Most bird gynandromorphs are bilaterally symmetrical, that is, the right half of the body is one sex and the left half is the other sex. Such individuals are rare but known to occur in several species of birds.

— H —

### Hallux
The hind toe of a bird's foot. The first of the four digits that form the foot.

### Hand
The portion of the wing extending outward from the wrist or "bend in the wing," also called a *manus*.

### Harderian glands
One of the lachrymal glands ("tear glands") associated with the eyes. Larger in marine birds where the rather oily secretions help protect the eyes from being damaged by the salt water.

### Hatching muscle
A muscle on the nape of a bird's neck prominent at the time of hatching but which degenerates shortly thereafter. Contraction of this muscle pulls the bird's head back and causes the beak to tap against the shell, helping fracture it to allow hatching to proceed.

### Hawking
A term for the behavior of a bird snatching an insect from midair.

### Haematozoan
A parasite which lives in the blood.

### Herbivorus
Having a diet of seed, buds and other plant parts.

### Hermaphrodite
An individual having both male and female sex organs. In most, the body type outwardly appears to be of one sex, but perhaps somewhat atypical. Birds showing both body types are even rarer (see Gynandromorph).

### Heterochrosis
Having abnormal coloration or plumage.

### Heterodactylous
Refers to the foot structure found in trogons where the first and second toes point rearward and the third and fourth toes point forward (see zygodactylous).

### Hippoboscid Flies
Parasitic flies who, as adults, live on and suck the blood of either mammals or birds (but not both). Over 90 species of these flies are known to parasitize birds. Once on an appropriate host, the adult flies do not leave.

### Homoiothermous
Having the ability to maintain a constant body temperature independent of the surrounding air temperature. "Warm-blooded."

### Humerus
The long bone of the upper wing (closest to the body).

### Hyoid bones
The bony structures which support the tongue.

### Hypophysis
The pituitary gland.

### Hypoptilum
Synonym for afterfeather.

### Hyporachis
An afterfeather's shaft.

## Ornithological Dictionary

## — I —

### Incubation patch
See Brood patch.

### Incumbent
Condition of the feet found in most passerines where the rearward pointing toe (hallux) is in the same plane as the forward pointing toes.

### Indeterminate egg-laying
Term used for birds which do not lay a fixed number of eggs per clutch.

### Infundibulum
The funnel-like end of the oviduct adjacent to the ovary.

### Innate behavior
Instinctive or inborn behaviors that are not learned.

### Insectivorous
Having a diet that is mainly insects.

### Interference colors
An iridescent effect caused by an interference of light rays when the feathers are turned at different angles relative to the viewer. (See iridescence.)

### Interspecific
Pertains to a interaction between two different species.

### Intraspecific
Pertains to interaction between two or more individuals of the same species.

### Iridescence
The effect of "shining" colors that are produced by the directional scattering of light rays reflected from the surface of special iridescent feathers. Hummingbirds often have iridescent feathers on the throat or on the back. The glossy black and purplish sheen on the head of the male Brewer's Blackbird is the result of iridescence.

## — J —

### Jugulum
The portion of the lower throat just above the breast.

### Juvenal
The first plumage of a bird that contains true contour feathers, acquired after the down is shed. Usually followed by the first adult plumage shortly after the bird leaves the nest but can last two to three months in some songbirds.

### Juvenile
A young bird that is out of the nest and mostly able to care for itself (though there may be some parental care), but not yet in full adult plumage.

## — K —

### Keel
A broad, flat, forward projecting extension of the sternum (breastbone) in birds where the flight muscles attach. Lacking in flightless birds, such as the ostrich, which have a flat sternum much more similar to mammals.

### Keratin
A substance composed of fine, microscopic fibers held together in a protein matrix. Feathers, claws, fingernails and hair are all made from keratin.

### Kleptoparasitism
Term used to describe the behavior of one species forcefully taking or "stealing" food from another species. Jaegers, for example, harass gulls to

force them to drop their food (usually fish or squid). When successful, the jaeger takes the dropped food.

## Kronism

European term for cannibalism upon one's own young (usually dead or weakened).

## — L —

## Lachrymal glands

The "tear glands" whose secretions help keep moisture over the eyes.

## Lamellate

Condition of some bird beaks, such as flamingos and some ducks, which have many fine serrations or lamellae along the edge of the bill. These lamellae act as a strainer for filtering out fine food particles from the water.

## Lappet

A fleshy, unfeathered, pendant growth of the fleshy part of the bill toward the rear of the mouth. Lappets may be found on wattlebirds of New Zealand (also see Dewlap and Wattle).

## Larynx

Thickened structure at the upper end of the trachea, the "voice box" of humans and other mammals. In birds, the larynx has no vocal cords but opens and closes the glottis which regulates airflow and prevents food or water from entering the trachea and respiratory system.

## Lek

An area where males of a bird species gather during the breeding season to display and attract females for mating. Mostly used in association with grouse, although some other species, notably the Ruff, also attend leks.

## Lobate

Refers to the condition of having lobed toes such as those found on grebes, coots and phalaropes. Each toe has a stiff fringe or webbing along the side but there is no webbing between the toes as in ducks or cormorants.

## Loomery

A breeding colony of murres.

## Loral stripe

"Eyestripe" extending only from the eye forward, not rearward from the eye. Yellow spots on a male Black-throated Gray Warbler are loral stripes.

## Lore

The small area on each side of a bird's face between the eyes and the base of the upper portion of the bill.

## — M —

## Malar stripe

The facial marking forming the "mustache" or cheek patch on each side of a bird's face. A malar stripe is especially prominent on the face of a male Northern Flicker.

## Mandible

The lower portion of a bird's beak.

## Mantle

The area of feathers covering the bird's back and top of its wings. Used especially when referring to gulls.

## Manus

The outer portion of the wing. Also known as the hand.

## Mast

Refers collectively to a variety of nuts (acorns, chestnuts, hazelnuts, etc.) and the seeds of coniferous trees.

## Maxilla
The upper half of a bird's beak.

## Meatus
Opening or passageway such as the auditory meatus which is the external ear canal.

## Melanin
Pigments of skin and feathers that are primarily responsible for dark colorations (mainly browns and black).

## Melanism or melanistic
Refers to a condition where excess melanin is produced (at least in some areas of the plumage), resulting in darker than normal coloration. Birds of northern latitudes often tend to be more melanistic than their more southern counterparts.

## Metacarpal
In vertebrate quadrupeds, the metacarpals are the bones of the hand between the wrist and the fingers. In birds, these are reduced to three (from five) and are fused to form the hand (manus) of the wing.

## Metatarsus
(Also called Tarsus or Tarsometatarsus.) The shank portion of a bird's foot between the ankle and the toes. In birds, all of the metatarsal bones have fused into one elongated bone. People often mistake this for the lower portion of the leg, but it is part of the foot.

## Mirror
Old term for the speculum on a duck's wing.

## Monogamous
Forming a pair-bond with only one mate during the breeding season.

## Monophagous
Having a diet that consists of only one type of food. Rare in birds. The Snail Kite is one example, as its sole diet is one species of snail.

## Monotypic species
A species that has only one race or subspecies. Osprey and Anna's Hummingbird are examples.

## Mutual preening
See Allopreening.

## Mycosis
General term for any disease caused by a fungus.

— N —

## Nail
The thickened, horny tip on the maxilla of geese, ducks and swans. Also called an unguis.

## Nape
The back of the neck.

## Nares
The external opening of the nostrils on a birds bill. (Naris is singular form.)

## Natal down
The first set of feathers to cover a young bird, acquired before hatching in precocial birds and after hatching in altricial birds.

## Nestling
Term used to describe a bird from the time it hatches until it leaves the nest.

## Nictitating membrane (nictitans)
The "third eyelid," a transparent inner eyelid that can cover the eye to moisten or protect it without blocking out light,

important to help prevent drying the surface of the eye during flight.

### Nidicole
"Nest-dweller." Term applied to the young of birds that remain in the nest for several days or weeks.

### Nidification
Nest building.

### Nidifugous
Birds whose young leave the nest soon after hatching.

### Nocturnal
Refers to animals whose main activities are carried out mostly at night.

### Nominate race or subspecies
Race or subspecies first used to describe and name the species as a whole.

### Nuchal
Of the nape.

### Nuptial plumage
The breeding plumage.

## — O —

### Obligate insectivore
A bird or other animal that requires insects as the main component of its diet.

### Obligate parasite
An organism which can survive only as a parasite on another species. The Brown-headed Cowbird is an obligate parasite as it never builds a nest or raises its own young.

### Occiput
Back portion of the head between the nape and the crown.

### Ocellated
Having "little eyes" (ocelli), or spots that resemble eyespots, such as the spots on a peacock's tail coverts.

### Oil gland
See Uropygial gland.

### Omnivorous
Having a diet that consists of both plant and animal material.

### Oology
The study of eggs.

### Orbit
The cavity in the skull where the eye is located.

### Ornithophilous
The term used for plants that are pollinated primarily or exclusively by birds.

### Oscines
Birds belonging to the suborder Passeri of the order Passeriformes. These are the "songbirds" (even though they do not all sing), defined as such based upon the structure of the syrinx and its associated muscles.

### Osteology
The study of bones.

### Otolith
The "ear stones" made of calcium carbonate, found in the inner ear.

### Ovary
The sex organ in the female where eggs are produced.

### Oviduct
The duct that carries the eggs released from the ovary to the uterus.

## Ornithological Dictionary

### Oviparous
Term for any animal which lays eggs. Live-bearing animals are termed viviparous. All birds are oviparous.

### Ovipositing
The act of laying one's eggs.

### Ovulation
The time when a mature egg ruptures from the ovary and is released into the oviduct.

### Palmate
Resembling a hand; in birds, term means having three toes in front fully webbed like ducks and geese.

### Pamprodactyl
Condition where all four toes of a bird's foot are turned forward, as in swifts.

### Parthenogenesis
Term for development of eggs without being fertilized by sperm. This is common in some insects but quite rare in vertebrates. Some unfertilized eggs may begin early development but soon abort and do not develop into adults. Some female turkeys can produce eggs that develop normally without being fertilized.

### Peep
A general term for any of the many small sandpipers which look much alike without careful examination.

### Pelagic
Term for species that spend most or all of their lives on or in the open ocean, e.g., albatrosses and shearwaters.

### Pen
A female swan.

### Pharynx
Throat cavity which leads into the esophagus and trachea.

### Pigeon's milk
Milky, protein-rich secretions of the crop of adult mated pigeons and doves fed to the young during the first few days after hatching.

### Pinfeather
A newly emerging feather which is still sheathed and not unfolded.

### Plumage
A collective term for all of the feathers on a bird.

### Plumules
Down feathers.

### Pollex
The first digit or thumb or the hand. See Alula.

### Polyandry
A breeding system in birds where one female has two or more males as mates. The Spotted Sandpiper is a polyandrous species.

### Polygamous
Having more than one mate.

### Polygyny
A breeding system where males mate with two or more females.

### Polymorphism
Pertaining to a species where there is more than one form or color phase. Snow Goose and "Blue Goose" are both members of the same polymorphic species (Snow Goose).

## Ornithological Dictionary

**Polytypic species**

A species where there is more than one race or subspecies.

**Postocular stripe**

An "eyestripe" found only behind the eye and not extending forward from the eye into the loral region.

**Powder down**

Highly modified feathers that grow continuously and the tips of which disintegrate into a fine waxy powder.

**Precocial**

Term used for birds that are hatched fully covered with down and with their eyes open. They do not remain in the nest after hatching.

**Preen gland**

See Uropygial gland.

**Preening**

The act of cleaning and reconditioning feathers. During preening, a bird cleans and smooths each feather, working oil from the uropygial gland through the plumage.

**Primary feather**

Large flight feathers attached to the hand portion (manus) of the wing.

**Promiscuous behavior**

Mating system in birds where no pair-bond is established. Males attempt to mate with as many females as possible. Exhibited by hummingbirds.

**Proventriculus**

Glandular stomach of birds. (The gizzard is muscular portion of stomach.)

**Proximal**

A point nearest or nearer to the point of reference. Opposite of distal.

**Pterylae**

Regions of the skin where feathers grow (feather tracts). Pterylae are separated from each other by open spaces called apteria.

**Pterylography**

The study of feather arrangement and growth on a bird.

**Pterylosis**

The arrangement of feathers in pterylae (feather tracts) and apteria.

**Pygostyle**

Tail (caudal) bone(s) where the tail feathers (rectrices) are attached; not a single bone but several vertebrae which have become modified and fused together during embryonic development.

**Pyloric stomach**

In some birds, this separate chamber is located between the gizzard and the small intestine. It may serve somewhat different functions for different groups of birds.

## — Q —

**Quill**

A term sometimes used to refer to a whole feather. The correct definition is more specific and refers only to the calamus (see calamus).

## — R —

**Race**

A subspecies.

**Rachis**

Portion of a feather shaft between vanes.

**Raptor**

Bird of prey that catches live food.

## Ratite
Ostrich, emu, cassowaries, kiwis and rheas are collectively known as ratites. All are flightless and lack a keel on the sternum.

## Reciprocal preening
See Allopreening.

## Rectrices
Large feathers of the tail, do not include the upper or lower tail coverts. (Singular: rectrix.)

## Recurved
Bill that is turned upward, like that of an avocet.

## Remiges
The flight feathers of the wings (primaries and/or secondaries). (Singular: remex.)

## Reproductive isolation
Condition of a species or subspecies unable to interbreed with another closely related species or subspecies. Florida Scrub-Jays (in Florida only) and the Island Scrub-Jay (only on Santa Cruz Island) are reproductively isolated from the Western Scrub-Jay (the most common species).

## Responsive singing
See Antiphonal singing.

## Reticulate
Scale pattern on the tarsus of a bird's foot where the scales appear in small polygonal patterns, found on an eagle.

## Rhamphotheca
Keratinized sheath covering the bill.

## Rictal bristle
Specialized, short, stiff barbless feathers (see Bristle), found around the corners of the mouth in swallows, flycatchers and nighthawks. May have some sensory function and assist with the ability to catch flying insects, although flycatchers have been shown to catch insects with only the tips of their bills.

## Rictus
Fleshy corners of the mouth at the rear of a bird's beak.

## Rookery
Originally a term for a nesting colony of Rooks. In the United States, it has a much broader meaning and is sometimes used synonymously with roost, although it is generally used to discuss nesting colonies of crows or herons.

## Roost
A place where birds (usually of the same or closely related species) flock together in trees, brush or tall grass to spend the night, especially during the non-breeding season.

## Roosting
Resting, sleeping at the roost location.

## Rostrum
Synonym of bill or beak.

— S —

## Salt gland
Specialized glands located above and behind the eyes that help a bird rid its system of excess salt. Such glands are totally absent in passerines and much reduced in most other land birds, but they are prominent in most seabirds. Without some way to get rid of excess salt, a bird

would not be able to drink saltwater. Especially well-developed in birds of the Order Procellariiformes (albatrosses, shearwaters, petrels and storm-petrels) which spend most or all of their life at sea.

## Scansorial
Adapted for climbing.

## Scapulars
A group of feathers at the bird's shoulders.

## Scleroid bones (sclerotic ring)
Series of thin, bony plates that form a ring overlying each eyeball; help to protect and strengthen the eyes.

## Scutellate
Pattern of scales on the tarsus of a bird's foot where scales overlap each other in the manner of shingles on a roof, such as is found on a starling.

## Scutellum
The scales or plates on the foot of a bird with a scutellate foot.

## Seasonal dimorphism
Reference to the pattern of many birds with two distinctly different plumages, one for the breeding season and one for the non-breeding season. This is the pattern, for example, of most male North American warblers.

## Shank
See Tarsus.

## Sibling species (Also known as cryptic species)
Two species so similar to one another throughout all seasons that they are nearly indistinguishable. One example is Willow and Alder Flycatchers, once considered to be one species known as "Traill's Flycatcher."

## Site tenacity
Refers to those birds that are faithful to their breeding grounds, returning in successive breeding seasons.

## Social parasitism
See Brood parasite.

## Social preening
See Allopreening.

## Speculum
A small patch of iridescent feathers, especially the area in the secondaries on the rear portion of a duck's wing.

## Squab
A young pigeon or dove.

## Stenophagous
A term used for birds that have a very restricted diet. Hummingbirds must eat nectar or small insects and mites. A monophagous diet is the most limited type of stenophagous behavior. Also see monophagous.

## Sternum
The breastbone. In flying birds, there is usually a strong keel on the sternum.

## Stomach oil
A strong and foul smelling oil regurgitated from the stomach of some birds, especially fulmars and other members of the Order Procellariiformes. Deliberately done as a method of self-defense when the birds are disturbed.

## Subadult
A young bird, not yet in adult plumage, of a species that requires more than one year to reach adulthood.

## Supercilium
The "eyebrow" or stripe above the eye of a bird.

## Supraorbital gland
See Salt gland.

## Sympatric species
Species which share much of the same range and habitat but are distinct species.

## Syndactyl
Foot pattern in birds in which toes are fused for much of their length, found in kingfishers and hornbills.

## Synsacrum
A series of solidly fused pelvic vertebrae in a bird. It extends from the last thoracic vertebra (these are the vertebrae from which the ribs extend), posteriorly to include the caudal or tail vertebrae.

## Syrinx
The voice-box of a bird. A structure unique to birds, located at the lower end of the windpipe at the point where the bronchi divides.

— T —

## Talon
Claw on the foot of a bird of prey.

## Tarsometatarsus
See Tarsus.

## Tarsus
Also known as the Metatarsus or Tarsometatarsus. The shank or portion of a bird's foot between the ankle and the toes. In birds, all of the metatarsal bones have become fused into one elongated bone. Many people mistake this for the lower portion of the leg but in reality, it is part of the foot.

## Tear gland
See Harderian glands.

## Teleoptiles
The mature feathers of an adult bird. Also called neossoptiles.

## Tertial or tertiary feather
The last three secondaries or innermost wing feathers. In some birds, they are distinct from the other secondaries.

## Testis
The male gonad responsible for production of sperm.

## Testosterone
The male sex hormone.

## Tibiotarsus
The major bone of the lower leg.

## Tomial tooth
A "tooth-like" projection on each side of the maxilla that is found in falcons and shrikes.

## Torpor
A relatively inactive state where physiological functions are greatly slowed but not as extreme as true hibernation. A torpid condition may last one to several hours. It is in a state of torpor that hummingbirds, with their very high metabolism and small bodies, survive cool nights.

## Totipalmate
A condition where all four toes of a bird's foot have a webbing between them. Cormorants and pelicans are totipalmate. Ducks and geese have only three toes that are connected by webbing and, therefore, are not totipalmate. Also see Palmate.

## Ornithological Dictionary

### Trachea
The air duct, surrounded by rings of cartilage and extending from the throat to the bronchi which branch toward the lungs.

### Tridactyl
Having only three toes.

### Trituration
The action taken by the gizzard of a bird, the grinding of hard foods.

### Type specimen
The actual specimen used to give the first scientific description for a species.

### Unguis
The nail on the tip of the upper bill found in ducks, geese and swans.

### Urodeum
The portion of the cloaca into which the genital ducts and ureters empty.

### Urohydrosis
Term to describe the action of some birds that urinate on their feet and legs in warm weather as a method of keeping cool. Since birds do not have sweat glands and cannot sweat, this is one method of evaporative cooling. It has been observed in storks, New World vultures and some cormorants.

### Uropygial gland
The "preen gland." Located just above the base of the tail, this gland has oil-rich secretions which a bird distributes throughout its plumage to help keep the feathers in good condition and prevent wetting. Birds usually use the top or back of their heads to reach this gland and distribute the oil.

### Uterus
In birds, the shell is added to the eggs in the uterus, at the end of the oviduct just before the vagina.

### Vagina
The vagina in female birds is at the end of the oviduct and is its most muscular portion. Mature eggs are quickly passed from this muscular vagina, through the cloaca and passed out of the body into the nest.

### Vane
The blade-like portion of a typical feather on either side of the rachis formed by the barbs and barbules.

### Vaned feathers
Any flight feather, tail feather or body (contour) feather which has a central shaft (rachis) and a stiff vane on either side.

### Vas deferens
The duct of the avian male reproductive system through which the sperm passes from the testis to the cloaca.

### Vent
The anus or opening of the cloaca.

### Ventral
Referring to the belly or underside.

### Ventriculus
The gizzard or muscular portions of a bird's stomach.

### Vestibule
Central chamber of the inner ear.

## — W —

### Wattle
The fleshy, unfeathered and often brightly colored pendant skin that hangs from the throat or lower bill (dewlap) or from the corners of the mouth (lappet).

### Wax-eating
See Cerophagy.

### Wing-loading
The ratio of a birds' weight to its wing surface area.

### Wingspan
The length of the wing.

## — X, Y, Z —

### Xanthochromism
A rare condition (except in some domesticated parrots) where little or no melanin is produced, allowing yellow carotenoids to predominate.

### Xenobiotics
Chemicals from outside the natural environment; may include insecticides, herbicides, industrial pollutants, heavy metals and a variety of other chemical substances. Many of these have serious negative impacts on birds and bird populations.

### Yolk
Membrane-bound portion of the egg containing protein-rich globules that provide nourishment to the developing embryo. Composed of two portions: the white yolk, deposited at night, and the yellow yolk, deposited during the day.

### Yoke-toed
See Zygodactyl.

### Zoonosis
A term for a disease that may be passed from one type of animal to another, including humans.

### Zygodactyl
Yoke-toed. The condition in bird's feet in which two toes point forward (2nd and 3rd) and two toes point rearward (1st and 4th). The feet of cuckoos, parrots and woodpeckers are zygodactyl.

### Zygote
The fertilized egg.

# Some Useful Books About Birds and Birding

American Ornithologists' Union. 1998. *Check-list of North American Birds.* 7th ed. American Ornithologists' Union, Washington, D.C.

Attenborough, David. 1998. *The Life of Birds.* Princeton University Press, Princeton, NJ.

Baicich, Paul J. and Colin J. O. Harrison. 1997. *A Guide To The Nests, Eggs, and Nestlings of North American Birds.* 2nd ed. Academic Press, San Diego, CA.

Brooks, Bruce. 1989. *On The Wing.* Charles Scribner's Sons, Macmillan Publishing Co., New York, NY.

Burton, Robert. 1992. *North American Birdfeeder Handbook.* National Audubon Society. DK Publishing, Inc. New York, NY.

Chatterjee, Sankar. 1997. *The Rise of Birds.* The Johns Hopkins University Press, Baltimore, MD.

Curson, Jon, David Quinn and David Beadle. 1994. *Warblers of the Americas: An Identification Guide.* Houghton Miflin Co. Boston, MA.

Dunn, Jon and Kimball Garrett. 1997. *A Field Guide to Warblers of North America.* Houghton Mifflin Co, Boston, MA.

Ehrlich, Paul R., David S. Dobkin and Darryl Wheye. 1988. *The Birder's Handbook.* Simon & Schuster, New York, NY.

Ehrlich, Paul R., David S. Dobkin and Darryl Wheye. 1992. *Birds in Jeopardy.* Stanford University Press, Stanford, CA.

Feduccia, Alan, 1996. *The Origin and Evolution of Birds.* Yale University Press, New Haven, CT.

Forshaw, Joseph, Steve Howell, Terence Lindsey, Rich Stallcup. 1995. *Birding.* The Nature Co., Time-Life Books

Gill, Frank B. 1990. *Ornithology,* 2nd ed. W.H. Freeman and Co., New York. NY.

Grant, P. J. 1997. *Gulls: A Guide to Identification,* 2nd ed. Academic Press, San Diego, CA.

Griggs, Jack L. 1997. *All the Birds of North America.* Harper Collins Publishers, New York, NY.

Haley, Delphine, ed. 1984. *Seabirds of Eastern North Pacific and Arctic Waters.* Pacific Search Press, Seattle, WA.

Harrison, Peter. 1983. *Seabirds: An Identification Guide.* Houghton Mifflin Co, Boston, MA.

Jobling, James A. 1991. *A Dictionary of Scientific Bird Names.* Oxford University Press, New York, NY.

Johnsgard, Paul A. 1975. *Waterfowl of North America.* Indiana University Press, Bloomington, IN.

Johnsgard, Paul A. 1988. *North American Owls: Biology and Natural History.* Smithsonian Institution Press, Washington D.C.

## Additional Reading

Kaufman, Kenn. 1996. *Lives of North American Birds*. Houghton Mifflin Company, Boston, MA.

Kaufman, Kenn. 1990. *Advanced Birding*. Houghton Mifflin Company, Boston, MA.

Kaufman, Kenn. 2000. *Birds of North America*. Houghton Mifflin Company, Boston, MA.

Kilham, Lawrence. 1989. *The American Crow and the Common Raven*. Texas A & M University Press, College Station, TX.

Kress, Stephen W. 1985. *The Audubon Society Guide to Attracting Birds*. Charles Scribner's Sons, New York, NY.

Lanner, Ronald, M. 1996. *Made for Each Other: A Symbiosis of Birds and Pines*. Oxford University Press, New York, NY.

Martin, Laura C. 1993. *The Folklore of Birds*. The Globe Pequot Press, Old Saybrook, CT.

Morris, Arthur. 1998. *The Art of Bird Photography*. Amphoto Books, New York, NY.

National Geographic Society. 1987. *Field Guide to Birds of North America* 3rd ed. National Geographic Society, Washington, D.C.

Paulson, Dennis. 1993. *Shorebirds of the Pacific Northwest*. University of Washington Press. Seattle Audubon Society, Seattle, WA.

Peterson, Roger Tory. 1979. *Penguins*. Houghton Mifflin Co, Boston. MA.

Peterson, Roger Tory. 1980. *A Field Guide to the Birds*. Houghton Mifflin Co, Boston. MA.

Peterson, Roger Tory. 1990. *A Field Guide to Western Birds*. Houghton Mifflin Co, Boston. MA.

Proctor, Noble S. and Partick J. Lynch. 1993. *Manual of Ornithology: Avian Structure & Function*. Yale University Press, New Haven, CT.

Savage, Candace. 1997. *Bird Brains: the Intelligence of Crows, Ravens, Magpies, and Jays*. Sierra Club Books, San Franciso, CA.

Sibley, David Allen. 2000. *The SIBLEY Guide to Birds*. Alfred A. Knopf, New york, N.Y.

Skutch, Alexander F. 1976. *Parent Birds and Their Young*. University of Texas Press, Austin, TX.

Skutch, Alexander F. 1989. *Birds Asleep*. University of Texas Press, Austin, TX.

Skutch, Alexander F. 1996. *The Minds of Birds*. Texas A&M University Press, College Station, TX.

Skutch, Alexander F. 1999. *Helpers at Birds' Nests: A Worldwide Survey of Cooperative Breeding and Related Behavior*. University of Iowa Press, Iowa City, IA.

Short, Lester L. 1993. *The Lives of Birds*. Henry Holt and Company, New York, NY.

Stokes, Donald and Lillian. 1996. *Stokes Field Guide to Birds: Eastern Region*. Little, Brown and Co., Boston, MA.

Stokes, Donald and Lillian. 1996. *Stokes Field Guide to Birds: Western Region*.

## Additional Reading

Little, Brown and Co., Boston, MA.

Terres, John K. 1980. *The Audubon Society Encyclopedia of North American Birds*. Alfred A. Knopf, New York, NY.

Tyrrell, Esther Quesada and Robert A. 1985. *Hummingbirds - Their Life and Behavior - A Photographic Study of the North American Species*. Crown Publishers, Inc., New York. NY.

Waldbauer, Gilbert. 1998. *The Birder's Bug Book*. Harvard University Press, Cambridge, MA.

Weidensaul, Scott. 1996. *Raptors: The Birds of Prey*. Lyons & Burford, New York, NY.

Weidensaul, Scott. 1999. *Living on the Wind: Across the Hemisphere with Migratory Birds*. North Point Press, New York, NY.

Weller, Milton W. 1999. *Wetland Birds: Habitat resources and Conservation Implications*. Cambridge University Press, Cambridge, U.K.

Welty, Joel Carl and Luis Baptista. 1988. *The Life of Birds*, 4th ed. Saunders College Publishing, New York, NY.

## Additional Reading

# Index

Illustrations are indicated with italics and "i" and photographs are indicated with a "p." This index contains references to the main body text of the book; the material in the appendixes has not been indexed.

— A —

Accentor, Hedge, 77, 80
Accipiter, 76
Acorn Woodpecker, 80
adaptations
  egg laying, 11, 92
  feeding, 53–57
  flight, 11–12
  skeletal, 21–22
*Aepyornis*, iii, 91
aerodynamics (flight), *12i–15i*, 12–16
African Weaver, 41
afterfeather (aftershaft), *2i*, 4
air-capillaries, 40
air currents, 16–17
air sacs, 22, 38–41, *39i*
albatross
  feathers, 4
  flight, 11, 15, *15i*, 16
  skeleton, 27-28
Alpine Chough, ii
alula, 14–15, *15i*, 29
American Bittern, *56i*, 60
American Coot, 32
American Crow, 36, 56–57
American Goldfinch, 84, 95, 102, *102i*
American Kestrel, 14, 19, 64
American Robin, 86, *87p*
American White Pelican, 7, 47, *47i*, *51i*
American Woodcock, 43, 60
Andean Hillstar, ii
angle of attack, *13i*, 14, *14i*
Ani, 4, 95
Anna's Hummingbird, 71, *93i*, 103, 110
apteria, 3
*Archaeopteryx*, 21
aspect ratios, 15-16
attack, angle of, *13i*, 14, *14i*
Audubon societies, iii

— B —

Bald Eagle, 2, *51i*
barbet, 86
Barn Owl, 67, 68, 96, 105
bastard wing, 29
bats, i, 21, 29, 69
Bay-winged Cowbird, 85

beaks. *See* bills
Bearded Vulture (Lammergier), 47
Bee Hummingbird, 103
Bernoulli, Daniel, 13
Bernoulli Effect, 13
bills, *23i*, 23–24, 43–45, *45i*, 70
Bird's-nest soup, 46
bird-watching, iii–iv
Bittern, American, *56i*, 60
Black-billed Magpie, 63
blackbirds, 79, 80, *92i*
Black Oystercatcher, *50i*, 55
Black Turnstone, 44, *49i*
blood, 33–37, *35i*
blood pressure, 34-36
Blue Tit, 54
bolus, 46
Boreal Owl, 68
botflies, 87–88
breathing rates, 40
breeding. *See* mating
bristle, *4i*, 6
bronchi, 38, 39, 40
brood parasitism, 83–89
Brown-headed Cowbird, 83, *84p*, 85
Brown Kiwi, 76
Brown Pelican, 41, *51i*
Brucke's muscle, 61, 63
Buff-breasted Sandpiper, 104–105
Bunting, Indigo, 87
Burrowing Owl, *98i*
Bustard, 46
Buteo, 14, *16i*
butterflies, Monarch, 71
Buzzard, Common, 64

— C —

caching, 55
calamus, 2
California Condor, 93
California Quail, 93–94
Canvasback, 89
Cape May Warbler, 100
capsaicins, 71
carina. *See* keel
carotenoids, 7, 8
Cave Swiftlet, 69

**Birds! From the Inside Out** 159

# Index

Cedar Waxing, 84
central streak, 65, 66
chalaza, 92, *94i*
Chickadee, 54, 101
chick
   cuckoo, 86
   feather growth, 9, *9i*
   feeding ("Pigeon's Milk"), 46
   hatching, 86, *95i*, 96–98
   parasitic, 84, 86–87
   vocalization, 87
chicken, 40, 70, 76
chord, 15
Chough, Alpine, ii
Chukar, *93i*
circulatory system, 33–37
Clark's Nutcracker, *55i*, 54–55
classification, 24
cloaca, 48
cloacal kiss, 77
cochlea, *67i*, 68
cold blooded. *See* poikilotherms
color, 7–9, 65–66
columella, *67i*, 68
Common Buzzard, 64
Common Cuckoo, 83, 85–86, *86i*
Common Murre, *93i*
Common Poorwill, *49i*, 104, *104i*
communication
   Black-billed Magpie, 63
   eyes, 62, 63
   head movement, 60
   parrot, 62
   *See also* vocalization
Condor, California, 93
cooling, evaporative, 40–41
coots, 32
cormorants, 47, *62i*
corvids, 56. *See also* Crows, Jays, Nutcracker
courtship. *See* mating
Cowbird
   Bay-winged, 85
   brood parasitism, 83–89
   Brown-headed, 83-85, *84p*
   Giant, 87-88
   Screaming, 85
Crampton's muscles, 61, 63
Crane, Sandhill, *78i*
crop, 46–47

crossbills, 47, 101, 109
Crow, American, 36, 56–57
Cuckoo, 30, 83, 85–86, *86i*
Curassow, 76
Curlew, Long-billed, 44

— D —

Dark-eyed Junco, 106, 109
depth of field, 63
digestive sytem, 12, 45–48, 71, 100
diseases
   aspergillosis, 112
   atherosclerosis, 36
   brood parasitism and, 87–88
distribution of habitat, ii–iii
diving birds
   cormorants, *62i*
   ears of, 67
   egg shape, *93i*
   nictating membrane of eye, 63
   underwater pressure and, 26
Domestic Turkey, 36, 40, 48, 76
dove
   incubation, 95
   homing ability, 69
   "Pigeon's Milk", 46
   survival in cold weather, 104
down feather, *3i*, 6
drag (flight), *12i–14i*, 12–15
ducks
   bill sensitivity, 43
   Black-headed, 88
   brood parasitism, 83, 88, 89
   Canvasback, 89
   Hooded Merganser, *49i*
   Mallard, ii, 43, 70
   mating system, 79
   migration, 106
   Northern Pintail, iv
   Northern Shoveler, 46
   Redhead, 88p, 89
   reproduction, 76, 77
   scent, 72
   tongue, 46
   Wood, *49i*
Dunlin, *49i*, 70, *70i*, 107
Dusky Moorhen, 80

— E —

Eagle
   Bald, 2, *51i*

## Index

eye size, 59
flight, 18
foot structure, 30
vision, 66
ears, 23i, 24, 66–69, 67i
echolocation, 69
Edible-nest Swiftlets, 46
eggs, 83–89, 91–97, 92i–95i. See also reproduction
egg tooth, 95i, 96–97
Egret, 6, 7, 57i
  Reddish, 57i
  Great, 7
Egyptian Vulture, 54
Elephant Bird, i, 91
Emu, 4, 21
esophagus, 46
eumelanin, 7
European Starling, 36, 72
evaporative cooling, 40
Evening Grosbeak, 43i, 45i, 113i
eyes, 22, 59–66, 61i

— F —

falcons
  American Kestrel, 14, 19, 64-65
  foot structure, 30
  reproduction, 76
  vision, 64-65, 66
feathers
  overview, 1–9
  color, 7-9
  contour, 2, 2i
  flight, 4, 5i, 15
  illustrations, 1i–5i, 9i
  photograph, 9p
  primary, 4, 5, 5i, 14, 15, 18
  vane, 2i, 3, 18
feather tracts, 1i
feeding. See also food
  adaptations, 43–51
  behaviors, 8, 53–57, 57i
  in flight, 17, 18
  hearing and, 68
  human participation in, 109–113
  mating and, 79, 95
  migration diet changes, 99–102, 107
  taste and smell and, 70–73
  vision and, 59, 66
feet, 29–32, 31i

filoplume, 4, 4i, 6
Finch
  bills, 45
  House, 8, 77, 109
  Mangrove, 53
  mate criteria, 77
  Purple, 77
  Rosy, 104
  Woodpecker, 53
Flamingo, 50i
flapping flight, 18-19
flight
  overview, 11–20
  energy requirements, 33
  feathers, 4, 5i, 15
  muscle size, 17, 25
  respiration and, 37
  weight reduction for, 11, 75
food. See also feeding
  butterflies, 71
  conifer buds, 101
  fish, 18
  flowers, 101
  fruit, 101
  horseshoe crab spawn, 108
  insects, 99
  leaves, 101
  mussels, 55
  nectar, 100, 111
  oils, 47-48
  seeds, 44, 45, 55
  sugar, 71, 100, 111
  sunflower seeds, 101
  voles, 66
  wax, 100
foot structure, 29–32
foraging. See feeding
forewing, 14, 15i, 29
Frigatebird, Magnificent, 21, 41

— G —

Galápagos Hawks, 80
Galápagos Islands, 53
game birds, 93i
geese, 7, 76, 79
Giant Cowbird, 87–88
gizzard, 47, 48
gland, preen (uropygial), 7, 25
gliding flight, 16–17
godwits, 70

## Index

Goldfinch, American, 84, 95, 102, *102i*
gonads, *75i*, 75–77, *76i*
Goose, Snow, 7
gravity (flight), *12i*, 12–14
Great Blue Heron, 12, *47i*, *93i*
Great Egret, 7
Greater Honeyguide, 86
Greater Sage-Grouse, 41, *41i*, *77i*, 78
Great Horned Owl, 22, 65, 68
Great Tit, 54, 54i
grebes, 32
Green Heron, *53i*, 54
Green Jay, 87
Grosbeak, Evening, *43i*, 44–45, *45i*, *113i*
ground effect, 17-18
Grouse
    Blue, 101, 101i
    Greater Sage, 41, *41i*, *77i*, 78
    mating, 81
    Ruffed, 103
gulls, 21, 36

— H —

habitat distribution, ii-iii
hallux, 29
hand, 14, 15
harrier, 76
Harris's Hawk, 80-81
hatching, *95i*, 96–98
Hawk
    Common Buzzard, 64
    eyesight, 64
    flight, 1-15
    foot structure, 30
    Galápagos, 80
    Harris's, 80
    Red-tailed, 14-15, 64-65, *64i*
    Sharp-shinned, 113
    vision, 64, 66
    wing shape, *16i*, 18
hearing, sense of, 66–69
heart, 33–37, *35i*
Hedge Accentor, 77, 80
hemoglobin, 34, 37
Herbst corpuscle, 69, *69i*, 70
Hermit, Little, 103
Hermit Thrush, 105–106
herons, 12, *47i*, *53i*, 54, *93i*
hibernation, 104
Hillstar, Andean, ii
Hoatzin, 47

hollow bones, 21–32
homeotherm (warm-blooded), 33
honeycreepers, 24
honeyguides, 83, 86
Hooded Merganser, *49i*
Horned Lark, ii
hovering flight, 19, *19i*
human
    bird-watching, iii-iv
    contrast (birds), 38, 59, 63, 70, 107
    fascination with birds, i
    feeding birds, 109–113
    flight inventions, 11
Hummingbird
    Andean Hillstar, ii
    Anna's, 71, *93i*, 103, 110
    Bee, 103
    breathing rates, 40
    competition, 71
    egg size, *93i*
    feathers, 4
    feeding, 70–71, 110, 111
    flight, 11, 19, *19i*
    heart rate, 36
    Little Hermit, 103
    mating, 81
    migration, 103, 107, 108, 110
    nectar eating, 100
    Ruby-throated, 2, 108
    Rufous, iii, *51i*, 71, *71i*, *81i*, 103, *103i*
    Scintillant, i
hunting. *See* feeding

— I —

identification of species, iii-iv
incubation, 94-96
Indigo Bunting, 87
infrasound, 69
infundibulum, *75i*, 77
intelligence, 53–57
intestines, 48

— J —

Jay
    coloration, 8
    food rejection, 71
    Green, 87
    saliva, 46
Junco, Dark-eyed, 106, 109

— K —

Kakapo, 47

## Index

keel, 21p, 25
keratin, 1, 9, 9i
Kestrel. See also falcon
   American, 14, 19, 64
   Eurasian, 66
Killdeer, 92i, 97i
kingfishers, 14, 19, 30, 63
King Penguin, 62
kites, 30
kittiwakes, 91
kiwis, 21, 72, 76, 91
Kiwi, brown, 76

— L —

Lammergeier (Bearded Vulture), 47
Lark, Horned, ii
leks, 77i, 81
lift (flight), 12i–14i, 12–16
Little Hermit, 103
Long-billed Curlew, 44
Long-eared Owl, 68
loons, 21, 32, 41
lungs, 38–40, 39i

— M —

Magnificent Frigatebird, 21, 41
Magpie, Black-billed, 63
Mallard, ii, 43, 70
mammals
   air flow, 38
   circulatory system, 33, 36
   digestive sytem, 12
   ears, 67, 68
   ruminants, 47
   skeleton, 24, 25, 26
   touch, 70
   vision, 60, 61
mandibles, 17–18, 23i, 23–24, 43
manus, 14, 15
Marsh Wren, 77-78, 79i, 80
mating
   anatomy, 75–77
   attracting mates, 77–79
   displays, 41, 77i, 81
   feeding mates, 95
   food abundance and, 101–102
   incubation, 94-96
   selecting mates, 77–81
maxilla. See mandibles
Meadowlark, Eastern and Western, ii
megapodes, 94, 97

melanin, 7
memory, 54-55
Merganser, Hooded, 49i
metabolic rates, 103
metabolism, 45
metacarpals, 28
migration, 99-101, 104-108
mitral valve, 34
molt, 8
monogamy, 79
Moorhen, Dusky, 80
Mousebird, 30
Murre, Common, 93i, 97
muscles
   bill, 43
   Brucke's, 61, 63
   Crampton's, 61, 63
   eye, 61
   flight, 27, 40
   pectoralis, 25, 27, 28, 102
   size, 17
   supracoracoideus, 28

— N —

Nashville Warbler, 100
nasofrontal hinge, 43
Native-Hen, Tasmanian, 80
nerve endings, 43
nests
   cliff, 91
   Edible-nest Swiftlet, 46
   lack of, 83
   material choices, 72
   mounds, 94
   reuse of, 104
   smell and, 72
nighthawks, 91
nocturnal birds, 65
Northern Pintail, ii
Northern Shoveler, 46
nostril, 40
nuthatches, 25, 104
Nuthatch
   Brown, 104
Nutcracker, Clark's, 55i, 55, 56

— O —

oilbirds, 69
oil droplets, 65
orbits, 23, 59
orioles, 86, 100

# Index

oropendolas, 87–88
Osprey, ii, 30, 67
ostrich
  eggs, 54, 91
  esophagus, 46
  skeleton, 21
  vision, 59
  weight, i
ovary, *75i*, 75, 76, 77
oviduct, *75i*, 75, 76, 77
Owls
  Barn, 67, 68, 96, 105
  Boreal, 68
  Burrowing, *98i*
  diurnal, 68
  ears, 67-68
  egg shape, *92i*
  eyes, 22, 59-61, 63
  foot structures, 30, *31i*
  Great Horned, 22, 65, 68
  hunting, 68
  incubation, 96
  Long-Eared, 68
  migration, 105
  skull shape, 68
  Snowy, 109
  vision, 60, 65
oxygen, 104
oystercatchers, *50i*, 55

— P —

pair-bond, 79, 81
parabronchi, 40
parasites, 78
parrots
  bills, 24, *50i*
  coloration, 8
  communication, 62
  foot structure, 30
patagium tendon, 19
pecten, 61, *61i*, 63, 64
pectoralis muscles, 25, 28, 102
Pelican
  American White, 7, *47i*, *51i*
  Brown, 41, *51i*
Penguin, King, 62
petrels, 48, 72
phaeomelanin, 7
phalaropes, 55, 80
pheasants, 55
pH levels, 37, 47

pigeons, 46
Pigeon's Milk, 46
pigments, 7–8
Pileated Woodpecker, *46i*
Pintail, Northern, ii
Plover
  bills, 44
  Black Oystercatcher, *50i*
  egg shape, *92i*
  Semipalmated, 44
poikilotherms (cold-blooded), 33
polyandry, 80
polygamy, 79
polygyny, 79
Poorwill, Common, *49i*, 104, *104i*
porphyrin, 7, 8
powderdown, 6
precocial, 97
predators, 45, 113
preen gland, *see* uropygial
preening, 6–7, 25
promiscuity, 81
ptarmigans, 103
pterygoid, 24, *25i*
pterylae, 1, 3
puffins, 23
pygostyle, 26

— Q —

Quail
  California, 93–94
  hatching patterns, 97
  incubation, 95

— R —

rachis, *2i*, 3-4, 18
rails, 80
ratites, 21. *See also individual species*
Réaumur, 48
rectrices, 4
Red Crossbill, *49i*, *102i*
Reddish Egrets, *57i*
Redhead, *88i*, 89
Red-tailed Hawk, 14-15, 64
Red-winged Blackbird, 80, *92i*
remiges, 4
reproductive system, 11–12, *75i*, 75–77, *76i*, 91–94. *See also* eggs
reptile comparisons
  circulatory system, 33, 36-37

# Index

skeleton, 23-24
vision, 61, 63
respiratory system, 12, 37–41, *39i*
rhamphotheca, 43
Robin, American, 86, *87i*
Ruby-throated Hummingbird, 2, 108
Rufous Hummingbird, ii, *51i*, 71, *71i*, 81, 103, *103i*

— S —

Sage-Grouse, Greater, 41, *41i*, *77i*, 78
saliva, 46
Sanderlings, 43, 70, *104i*, 105. See *also* shorebirds
Sandgrouse, ii
Sandhill Crane, *78i*
Sandpipers
　Black Turnstone, *49i*
　Buff-breasted, 104
　Dunlin, *49i*, 70, *70i*, 107
　food discrimination, 70
　foot structure, 30, *31i*
　mating system, 80
　migration, 105, 107
　Sanderlings, 43, 70, *104i*, 105
　Spotted, 80, *80i*, 95
　turnstones, 44
scapula, 27
scent, 72, 73
Scintillant Hummingbird, i
sclerotic ring, 23, 61, *61i*
screamers, 26
Screaming Cowbird, 85
second wing, 28-29
seed storage (caching), 55
semialtricial, 98
semiplume, *3i*, 6
semiprecocial, 98
senses, See individual senses
Sharp-shinned Hawk, 113
Shearwaters, 4, 17
shells. See eggs
shorebirds, 41, *43i*, 44, *49i*, 70, 107, 108
　See *also* sanderlings, sandpipers, plovers
Shoveler, Northern, 46
shrikes, 47
Skimmer
　Black, *50i*
　feeding behavior, 17, 18
　flight, 17–18, *18i*

habitat, 18
jaw structure, 24
pupil shape, 62
Skylark, ii
smell, sense of, 72–73
snipe, 4, 60, *60i*, 70
Snow Goose, 7
Snowy Owl, 109
soaring, 16–17
songbirds
　blood pressure, 36
　Chickadees, 101, *109i*
　crows, 36, 56
　cuckoos, 30, 83, 85, 86
　diet, 99–102
　Hedge Accentor, 77, 80
　Redpoll, 47
　shrike, 47
　Skylark, ii
　sparrows, 79, 85, 87, 105, 106, 109
　starlings, 36, 72
　swallows, 46, 78, 89, 104
　tanagers, 100
　thrushes, *87p*, 105–106, *106i*
　tits, European species, 54, *54i*
　towhees, 109
　vireos, 9, 100-101
　warblers, 99–100
　waxwings, 84
　wrens, iv, 77-78, *79i*, 80, 105
songs, 66–67
sounds, 69
Sparrows
　Chipping, 79
　Fox, 106, *107i*
　Golden-crowned, 109
　House, 105
　Song, 86-87
　White-crowned, *105i*, 105
Spotted Sandpiper, 80, *80i*, 95
Spotted Towhee, 109
stall, 14, 19
Starling, European, 36, 72
sternum, 22, 25, 38
stomach, 47–48
sugar and hummingbirds, 71
superprecocial, 97
supracoracoideus muscle, 28
Swainson's Hawk, *64i*
Swainson's Thrushes, 105–106, *106i*

**Birds! From the Inside Out**　　　　　　　　　165

## Index

Swallow
  Barn, 78
  brood parasitism, 89
  Cliff, *88p*, 89, *89i*, 104
  diet, 99
  mate selection, 78, 79
  nesting habits, 104
  saliva, 46
  Tree, 79
  vision, 59
swans, 2, 25
sweat glands, 112
swiftlet
  cave, 69
  Edible-nest, 46
Swifts
  diet, 99
  foot structure, 30
  mating, 77
  saliva, 46
  vision, 59
  wings, 28

— T —

tails, 19–20, 78
tanagers, 100
Tasmanian Native-hen, 80
taste, sense of, 70–72
temperature regulation, 102–104
tendons, 19
Tennessee Warbler, 100
testes, 76, *76i*
territories, 79
Thrush
  Hermit, 105–106
  Swainson's, 105–106, *106i*
thrust (flight), 12, *12i*, 13, *13i*, 15
Tit, 54, *54i*
  Blue, 54
  Great, 54, *54i*
tongues, 46
tool-using feeding behaviors, 53, 54
torpor, 103, 104
touch, sense of, 69–70
Towhee, Spotted, 109
Townsend's Warbler, 100, *100i*
tropicbirds, 11
tropics, ii–iii
troupials, 87
Tundra Swan, 2
turbulence in flight, 13, 14, 15

Turkey, Domestic, 36, 40, 48, 76
Turkey Vulture, 17, *72i*, 72–73
Turnstone, Black, *49i*
turnstone, 44

— U —

ultrasonic sound, 69
ultraviolet light, 66
uncinate process, 26
uric acid, 12
urine, 12
uropygial (preen) gland, 7, 25
uterus, 75, *75i*

— V —

vagina, 75, *75i*
velocity gradient, 17
vertebrae. *See* skeleton
vireos, 9, 100
vision, sense of, 59–66
vocalization, 59
vulture flight, 17
Vultures
  Bearded (Lammergier), 47
  Egyptian, 54
  Turkey, 17, *72i*, 72–73

— W —

warblers, 99–100
Warbler
  Nashville, 100
  Tennessee, 100
  Townsend's, 100, *100i*
  Yellow, *51i*, 84, 85p, 86–87
warm-blooded. *See* homeotherm
wax-eating, 99–100
Waxwing, Cedar, 84
Weaver, African, 41
weight and flight, *12i*, 12–13
Whydah, 85
wing-loading, 15–16
wings, i, 12–16, *13i–15i*, 26, 27–29
wingspan, 15
Winter Wren, ii, 105
Woodcock, American, 43, 60
Woodpeckers
  Acorn, 80
  brood parasitism of, 86
  carina, 25
  foot structures, 30
  incubation, 95
  nostril protection, 40

## Index

Pileated, *46i*
  saliva, 46
  tongue use, 70
wrens, ii, *79i*, 80, 105
wrists, *5i*, 18, 19

— Y —

Yellow-headed Blackbird, 80
Yellow-rumped Warblers, 99–100
Yellow Warbler, *51i*, 84, *85p*, 86–87
yolk, 92, *94i*

## Dan and Barbara Gleason

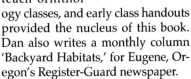

Dan and Barbara Gleason are a natural science writer-photographer and artist-illustrator who love working, birding and exploring the outdoors. They teach ornithology classes, and early class handouts provided the nucleus of this book. Dan also writes a monthly column 'Backyard Habitats,' for Eugene, Oregon's Register-Guard newspaper.

Dan worked in the Biology Department at the University of Oregon until retiring and he continues to teach Field Ornithology each summer. He has taught workshops at the Malheur Field Station, Sitka Center for Art & Ecology, and other centers, and he has developed and taught docent trainings on natural history and bird biology. He is a popular invited speaker at bird conferences and festivals, and for Audubon chapters and bird- and nature-oriented organizations.

Dan served as a director on the board of directors of the Lane County Audubon Society and Oregon Field Ornithologists, and he volunteers at Cascades Raptor Center, providing instruction about birds to staff, volunteers and the public. He enjoys retirement by writing, photographing and exploring the outdoors, as well as restoring old photographs.

Barbara is a colored pencil artist, natural science illustrator and graphic designer who manages BGleason Design & Illustration, and CraneDance Publications, their book-publishing wing. She provides design, illustration and consulting services to clients that range from individual authors to numerous companies and government agencies.

A member of the Guild of Natural Science Illustrators and the Colored Pencil Society of America, she has also been active in a number of business and art groups. She served on the board of directors of Lane County Audubon Society and is the corporate image director for the Colored Pencil Society of America.

Barbara paints birds, wildlife and landscapes and helps maintain a bevy of wild bird feeders in order to study her subjects better.

Together, Barbara and Dan teach Nordic Walking and they enjoy many activities outdoors in nature. They live in Eugene, Oregon, with two bird-watching cats who enjoy watching birds (including turkeys), squirrels and deer from their enclosed "catio" and *all* enjoy the many outdoor birds at their feeders.

---

The Gleasons can be reached by email at info@bgleasondesign.com, by mail at P.O. Box 50535, Eugene OR 97405, or by phone at (541) 345-3974.